Applied Data Mining

Applied Data Mining

Guandong Xu
University of Technology Sydney
Sydney, Australia

Yu Zong
West Anhui University
Luan, China

Zhenglu Yang
The University of Tokyo
Tokyo, Japan

CRC Press
Taylor & Francis Group
Boca Raton London New York

CRC Press is an imprint of the
Taylor & Francis Group, an **informa** business

A SCIENCE PUBLISHERS BOOK

CRC Press
Taylor & Francis Group
6000 Broken Sound Parkway NW, Suite 300
Boca Raton, FL 33487-2742

© 2013 Copyright reserved
CRC Press is an imprint of Taylor & Francis Group, an Informa business

International Standard Book Number: 978-1-4665-8583-6 (Hardback)

Visit the Taylor & Francis Web site at
http://www.taylorandfrancis.com

CRC Press Web site at
http://www.crcpress.com

Science Publishers Web site at
http://www.scipub.net

Preface

The data era is here. It provides a wealth of opportunities, but also poses challenges for the effective and efficient utilization of the huge data. Data mining research is necessary to derive useful information from large data. The book reviews applied data mining from theoretical basis to practical applications.

The book consists of three main parts: Fundamentals, Advanced Data Mining, and Emerging Applications. In the first part, the authors first introduce and review the fundamental concepts and mathematical models which are commonly used in data mining. There are five chapters in this section, which lay a solid base and prepare the necessary skills and approaches for further understanding the remaining parts of the book. The second part comprises three chapters and addresses the topics of advanced clustering, multi-label classification, and privacy preserving, which are all hot topics in applied data mining. In the final part, the authors present some recent emerging applications of applied data mining, i.e., data stream, recommender systems, and social tagging annotation systems. This part introduces the contents in a sequence of theoretical background, state-of-the-art techniques, application cases, and future research directions.

This book combines the fundamental concepts, models, and algorithms in the data mining domain together, to serve as a reference for researchers and practitioners from as diverse backgrounds as computer science, machine learning, information systems, artificial intelligence, statistics, operational science, business intelligence as well as social science disciplines. Furthermore, this book provides a compilation and summarization for disseminating and reviewing the recent emerging advances in a variety of data mining application arenas, such as advanced data mining, analytics, internet computing, recommender systems as well as social computing and applied informatics from the perspective of developmental practice for emerging research and practical applications. This book will also be useful as a textbook for postgraduate students and senior undergraduate students in related areas.

This book features the following topics:

- Systematically presents and discusses the mathematical background and representative algorithms for data mining, information retrieval, and internet computing.
- Thoroughly reviews the related studies and outcomes conducted on the addressed topics.
- Substantially demonstrates various important applications in the areas of classical data mining, advanced data mining, and emerging research topics such as stream data mining, recommender systems, social computing.
- Heuristically outlines the open research issues of interdisciplinary research topics, and identifies several future research directions that readers may be interested in.

April 2013

Guandong Xu
Yu Zong
Zhenglu Yang

Contents

Part II: Advanced Data Mining

Part III: Emerging Applications

Part I

Fundamentals

CHAPTER 1

Introduction

In the last couple of decades, we have witnessed a significant increase in the volume of data in our daily life—there is data available for almost all aspects of life. Almost every individual, company and organization has created and can access a large amount of data and information recording the historical activities of themselves when they are interacting with the surrounding world. This kind of data and information helps to provide the analytical sources to reveal the evolution of important objects or trends, which will greatly help the growth and development of business and economy. However, due to the bottleneck of technological advance and application, such potential has yet been fully addressed and exploited in theory as well as in real world applications. Undoubtedly, data mining is a very important and active topic since it was coined in the 1990s, and many algorithmic and theoretical breakthroughs have been achieved as a result of synthesized efforts of multiple domains, such as database, machine learning, statistics, information retrieval and information systems. Recently, there has been an increasing focus shift in data mining from algorithmic innovations to application and marketing driven issues, i.e., due to the increasing demand from industry and business, more and more people pay attention to applied data mining. This book aims at creating a bridge between data mining algorithms and applications, especially the newly emerging topics of applied data mining. In this chapter, we first review the related concepts and techniques involved in data mining research and applications. The layout of this book is then described from three perspectives—fundamentals, advanced data mining and emerging applications. Finally the readership of this book and its purpose is discussed.

1.1 Background

We are often overwhelmed with various kinds of data which comes from the pervasive use of electronic equipment and computing facilities, and whose size is continuously increasing. Personal computing devices are becoming cheap and convenient, so it is easy to use it in almost every aspect of our daily life, ranging from entertainment and communication to education and political life. The dropping down of prices of electronic storage drivers allows us to purchase disks to save information easily, which had to be discarded earlier due to the expense reason. Nowadays database and information systems have been widely deployed in industry and business, and they have the capability to record the interactions between users and systems, such as online shoppings, banking transactions, financial decisions and so on. The interactions between users and database systems form an important data source for business analysis and business intelligence. To deal with the overload of information, search engines have been invented as a useful tool to help us locate and retrieve the needed information over the Internet. The user navigational and retrieval activities that have been recorded in Web log servers, undoubtedly can convey the browsing behavior and hidden intent of users that are explicitly unseen, without in-depth analysis. Thus, the widespread use of high-speed telecommunication infrastructures, the easy affordability of data storage equipment, the ubiquitous deployment of information systems and advanced data analysis techniques have put us in front of an unprecedented data-intensive and data-centric world. We are facing an urgent challenge in dealing with the growing gap between data generation and our understanding capability. Due to the restricted volume of human brain cells, an individual's reasoning, summarizing and analyses is limited. On the contrary, with the increase in data volume, the proportion of data that people can understand decreases. These two facts bring a real demand to tackle the realistic problem in current information society—it is almost impossible to simply rely on human labors to accomplish the data analysis more scalable and intelligent computational methods are called for urgently. *Data mining* is emerging as one kind of such technical solutions to address these challenges and demands.

1.1.1 Data Mining—Definitions and Concepts

Data mining is actually an analytical process to reveal the patterns or trends hidden in the vast data ocean of data via cutting-edge computational intelligence paradigms [5]. The original meaning of "mining" represents the operation of extracting precious resources such as oil or gold from the earth. The combination of mining with the word "data" reflects the in-depth analysis of data to reveal the knowledge "nuggets" that are not exposed explicitly in the mass of data. As the undiscovered knowledge is of statistical nature, via statistical means, it is sometimes called statistical analysis, or multivariate statistical analysis due to its multivariate nature. From the perspective of scientific research, data mining is closely related to many other disciplines, such as machine learning, database, statistics, data analytics, operational research, decision support, information systems, information retrieval and so on. For example, from the viewpoint of data itself, data mining is a variant discipline of database systems, following research directions, such as data warehousing (on storage and retrieval) and clustering (data coherence and performance). In terms of methodologies and tools, data mining could be considered as the sub-stream of machine learning and statistics—revealing the statistical characteristics of data occurrences and distributions via computational or artificial intelligence paradigms.

Thus data mining is defined as the process of using one or more computational learning techniques to analyze and extract useful knowledge from data in databases. The aim of data mining is to reveal trends and patterns hidden in data. Hence from this viewpoint, this procedure is very relevant to the term *Pattern Recognition*, which is a traditional and active topic in *Artificial Intelligence*. The emergence of data mining is closely related to the research advances in database systems in computer science, especially the evolution and organization of databases, and later incorporating more computational learning approaches. The very basic database operations such as query and reporting simulate the very early stages of data mining. Query and reporting are very functional tools to help us locate and identify the requested data records within the database at various granularity levels, and present more informative characteristics of the identified data, such as statistical results. The operations could be done locally and remotely, where the former is executed at local end-user side, while the latter over a distributed network environment, such as the Intranet or Internet. Data retrieval, similar to data mining, extracts the needed data and information from databases. In order to filter out the needed data from the whole data repository, the database administrators or end-users need to define beforehand a set of constraints or filters which will be employed at a later stage. A typical example is the marketing investigation of customer groups who have bought two products consequently by using the "and" joint operator to form a filter, in order to identify the specific customer group. This is viewed as a simplest business means in marketing campaign. Apparently, the database itself offers somewhat surface methods for data analysis and business intelligence but far from the real business requirements such as customer behavioral modeling and product targeting.

Data mining is different from data query and retrieval because it drills down the in-depth associations and coherences between the data occurrence within the repository that are impossible to be known beforehand or via using basic data manipulating. Instead of query and retrieval operations, data mining usually utilizes more complicated and intelligent data analysis approaches, which are "borrowed" from the relevant research domains such as machine learning and artificial intelligence. Additionally, it also allows the supportive decision made upon the judgment on the data itself, and the knowledgeable patterns derived. A similar data analytical method is called *Online Analytical Processing* (**OLAP**), which is actually a graphic data reporting tool to visualize the multidimensional structure within the database. OLAP is used to summarize and demonstrate the relations between available variables in the form of a two-dimensional table. Different from OLAP, data mining brings together all the attributes and treats them in a unified manner, revealing the underlying models or patterns for real applications, such as business analytics. In one word, OLAP is more like a visualization instrument, whereas, data mining reflects the analytical capability for more intelligent use. Although data query, retrieval and OLAP and data mining have owned a lot of commonplaces, data mining is distinctive from the counterparts due to its outstanding and competent advantages of analysis.

Knowledge Discovery in Database (**KDD**) is a name frequently used interchangeably together with data mining. In fact, data mining has a broader coverage of applicability while KDD is more focused on the extension of scientific methods in data mining. In addition to performing data mining, a typical KDD process also includes the stages of data collection, data preprocessing and knowledge utilization, which form a whole cycle of data preparation, data mining or knowledge discovery and knowledge utilization. However it is indeed hard to draw a clear border to differentiate these two kinds of disciplines since there is a big overlapping between the two from the perspectives of not only the research targets and approaches, but also the research communities and publications. More theoretically, data mining is more about data objects and algorithms involved, while KDD is a synergy of knowledge discovery process and learning approaches used. In this book, we mainly focus our description on data mining, presenting a generic and broad landscape to bridge the gap between theory and application.

1.1.2 Data Mining Process

The key components within a data mining task consist of the following subtasks:

- Definition of the data analytical purposes and application domain.
- Data organization and design structure, data preparation, consolidation and integration.
- Exploratory analysis of the data and summarization of the preliminary results.

- Computational learning approach choosing and devising based on data analytical purposes.
- Data mining process using the above approaches.
- Knowledge representation of results in the form of models or patterns.
- Interpretation of knowledge patterns and the subsequent utilization in decision supports.

1.1.2.1 Definition of Aims

Definition of aims is to clearly specify the analytical purpose of data mining, i.e., what kinds of data mining tasks are intended to be conducted, what major outcomes would be discovered, what the application domain of the data mining task is, and how the findings are interpreted based on domain expertise. A clear statement of the problem and the aims to be achieved are the prerequisite for setting up the mining task correctly and the key for fulfilling the aims successfully. The definition of the analytical aims also prepares a guidance for the data organization and the engaged data mining approaches in the following subtasks:

1.1.2.2 Design of Data Schema

This step is to design the data organization upon which the data analysis will be performed. Normally in a data analysis task, there are a handful of features involved, and these features can be accommodated into various data models. Hence choosing an appropriate data schema and selecting the related attributes in the chosen schema is also a crucial procedure in the success of data mining. Mathematically, there exist some well studied models, such as *Vector Space Model* (VSM) and *graph* model to choose from. We need to choose a practical model to reflect and accommodate the engaged features. Features are another important consideration in data mining, which is used to describe the data objects and characterize the individual property of the data. For example, given a scenario of customer credit assessment in banking applications, the considered attributes could include customers' age, education background, salary income, asset amount, historic default records and so on. To induce the practical credit assessment rules or patterns, we need to carefully select the possibly relevant attributes to form the features of the chosen model. There are a number of feature selection algorithms developed in past studies of data mining and machine learning. An additional concern is the diverse residency of data in multiple databases due to the current distributed computing environment and popularization of internal or external networking. In other words, the selected data attributes are distributed in different databases locally and remotely. Thus data federation and consolidation is often a necessary step to deal with the heterogeneity and homogeneity of multiple databases. All these operations comprise the data preparation and preprocessing of data mining.

1.1.2.3 Exploratory Analysis

Exploratory analysis of the data is the process of exploring the basic statistical property of the data involved. The aim of this preliminary analysis is to transform the original data distribution to a new visualization form, which can be better understood. This step provides the start to choose appropriate data mining algorithms since the suitability of various algorithms is largely dependent on the data integrity and coherence. The exploratory analysis of the data is also able to identify the anomalous data—the entries which exhibit distinctive distribution or occurrence, sometimes also called outliers, and the missing data. This can trigger the additional data preprocessing operations to assure the data integrity and quality. Another purpose of this step is to suggest the need for extraction of additional data since the obtained data is not rich enough to conduct the desired tasks. In short, this stage works as a prerequisite to connect the analytical aims and data mining algorithms, facilitating the analytical tasks and saving the computational overhead for algorithm design and refinement.

1.1.2.4 Algorithm Design and Implementation

Data mining algorithm design and implementation is always the most important part in the whole data mining process. As discussed above, the selection of appropriate analytical algorithms is closely related to the analytical purposes, the organization of data, the model of analysis task and the initial exploratory analysis on the constructed data source. There is a wide spectrum of data mining algorithms that can be used to tackle the requested tasks, so it is essential to carefully select the appropriate algorithms. The choice of data mining algorithms are mainly dependent on the used data itself and the nature of the analytical task. Benefiting from the advances and achievements in related research communities, such as machine learning, computational intelligence and statistics, many practical and effective paradigms have been devised and employed in a variety of applications, and great successes have been made. We can categorize these methods into the following approaches:

- *Descriptive approach*: This kind of approach aims at giving a descriptive statement on the data we are analyzing. To do this, we have to look deeply into the distribution of the data, reveal the mutual relations among the objects, and capture the common characteristics of data distribution via machine intelligence methods. For example, clustering analysis is used to partition data objects into various groups unknown beforehand based on the mutual distance or similarity between them. The criterion of such partition is to meet the optimal condition that the objects within the same group are close to each other, while the objects from different groups should be separated far enough. Topic modeling is a newly emerging descriptive learning method to detect the topical coherence with the observations. Through the adjustment of the statistical model chosen for learning and comparison between the observation and model derivation, we can identify the hidden topic distribution underlying the observations and associations between the topics and the data objects. In this way all the objects are treated equally and an overall and statistical description is derived from the machine learning process. As they mainly rely on the computational power of machines without human interactions, sometimes we also call them unsupervised approaches.

- *Predictive approach*: This kind of approach aims at concluding some operational rules or regulations for prediction. By generalizing the linkage between the outcome and observed variables, we can induce some rules or patterns of classifications and predictions. These rules help us to predict the unknown status of new targeted objects or occurrence of specific results. To accomplish this, we have to collect sufficient data samples in advance, which have been already labeled with the specific input labels, for example, the positive or negative in pathological examination or accept and reject decision in bank credit assessment. These approaches are mainly developed in the domain of machine learning such as *Support Vector Machine* (SVM), decision tree and so on. The learned results from such approaches are represented as a set of reasoning conditions and stored as rule to guide the future prediction and judgment. One distinct feature of this kind approaches is the presence of labeled samples beforehand and the classifier are trained upon the training data, so it is also called supervised approaches (i.e., with prior knowledge and human supervision). Predictive approaches account for majority of analytical tasks in real applications due to its advantage for future prediction.

- *Evolutionary approach*: The above two kinds of approaches are often used to deal with the static data, i.e., data collected is restricted within a specific time frame. However, with the huge reflux of massive data available in a distributed and networked environment, the dynamics becomes a challenging characteristic in data mining research. This calls for evolutionary data mining algorithms to deal with the change of temporal and spatial data within the database. The representative methods and applications include sequential pattern mining and data stream mining. The former is to determine the significant patterns from the sequential data observations, such as the customer behavior in online shopping, whereas the latter was proposed to tackle the difficulties within data stream applications, such as RFID signal sampling and processing. The main difference of this with other approaches is the outstanding capability to deal with continuous signal generating and processing in real time with affordable computational cost, such as limited memory and CPU usage. Recently, such approaches highlight this new active and potential trends within data mining research.

- *Detective approach*: the descriptive and predictive approaches are focused on the exploration of the global property of data rather than that of local information. Sometimes the analysis at the smaller granularity will provide us more informative findings than the overall description or prediction. Detective approaches are the means to help us uncover the local mutual relations at a lower level. In data mining, association rule mining or sequential pattern mining are able to fulfill such requirement within a specific application domain, such as business transaction data or online shopping data.

Although four categories from the perspectives of data objects and analysis aims are presented, it is worth noting that the dividing lines between all these approaches are blurred and overlap one other. In real applications, we often take a mixture of these approaches to satisfy the requirements of complexity and practicality. More often, using the existing approaches or a mixture of them is a far cry from the success of analytical tasks in real applications, resulting in the desire to design new innovative algorithms and implementing them in real scenarios with satisfactory performance. This inspires researchers from different communities to make more efforts and fully utilize the findings from relevant areas.

Another significant issue attracting our attention is the increasingly popularity of data mining in almost every aspect of business, industry and society. The real analytical questions have raised a bunch of new challenges and opportunities for researchers to form the synergy to undertake applied data mining, which lays down a solid foundation and a real motivation for this new book.

1.1.3 Data Mining Algorithms

1.1.3.1 Descriptive and Predictive

Due to the broad applications and unique intelligent capability of data mining, a huge amount of research efforts have been invested and a wide spectrum of algorithms and techniques have been developed [5]. In general, from the perspective of data mining aims, data mining algorithms can be categorized into two main streams: descriptive and predictive algorithms. Descriptive approaches aim to reveal the characteristic data structure hidden in the data collection, while the predictive methods build up prediction models to forecast the potential attribute of new data subjects instead.

There are various descriptive data mining approaches that have been devised in the past decades, such as data characterization, discrimination, association rule mining, clustering and so on. The common capability of such kinds of approaches is to present the data property and describe the data distribution in a mathematical manner, which is not easily seen at surface analysis. Clustering is a typical descriptive algorithm, indicating the aggregation behavior of data objects. By defining the specific distance or similarity measure, we are able to capture the mutual distance or similarity between different data points (as shown in Fig.1.1.1). In contrast, predictive approaches mainly exploit the prior knowledge, such as known labels or categories, to derive a prediction "model" that best describes and differentiates data classes. As the model is learned from the available dataset by using machine learning approaches, the process is also called model training, while the dataset used is therefore named training data (i.e., data objects whose class label is known). After the model is trained, it is used to predict the class label for new data subjects based on the actual attribute of the data.

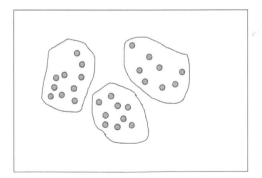

Figure 1.1.1: Cluster analysis

1.1.3.2 Association Rule and Frequent Pattern Mining

Association rule mining [1] is one of the most important techniques in the data mining domain, which is to reveal the co-occurrence relationships of activities or observations in a large database or data repository. Suppose in a traditional e-marketing application, the purchase consequence of "milk" and "bread" is a commonly observed pattern in any supermarket case, therefore resulting the generating of association rule ⟨bread, milk⟩. Of course, there may exist a large number of association rules in a huge transaction database dependent on the setting of the satisfactory (or confidence) threshold. The algorithm of association rule mining is thus designed to extract such rules as are hidden in the massive data based on the analyst's targets. Figure 1.1.2 gives a typical association rule set in a market-basket transaction campaign. Here you can observe the common occurrence of various items in supermarket transaction records, which can be used to improve the market profit by adjusting the item-shelf arrangement in daily supermarket management. Frequent pattern mining is one of the most fundamental research issues in data mining, which aims to mine useful information from huge volumes of data [4]. The purpose of searching such frequent patterns (i.e., association rules) is to explore the historical supermarket transaction data, which is indeed to discover the customer behavior based on the purchased items.

TID	Items
1	Bread, Milk
2	Bread, Diaper, Beer, Eggs
3	Milk, Diaper, Beer, Coke
4	Bread, Milk, Diaper, Beer
5	Bread, Milk, Diaper, Coke

Figure 1.1.2: An example of association rules

1.1.3.3 Clustering

Clustering is an approach to reveal the group coherence of data points and capture the partition of data points [2]. The outcome of clustering operation is a set of clusters, in which the data points within the same cluster have a minimum mutual distance, while the data points belonging to different clusters are sufficiently separated from each other. Since clustering is performed relying on the data distribution itself, i.e., the mutual distance, but not associated with other prior knowledge, it is also called *unsupervised algorithm*. Figure 1.1.3 depicts an example of cluster analysis of debt-income relationships.

1.1.3.4 Classification and Prediction

Classification is a typical predictive method. The aim of classification is to determine the class (or category) label for data objects based on the trained model (sometimes also called classifier). It is hard to completely differentiate the prediction approach from classification. In the data mining community, one commonly agreed opinion is that classification is mainly focused on determining the categorical attribute of data objects, while prediction is focused on continuous-values attributes instead, i.e., it is used to predict the analog values of data objects. As the model learning and prediction is performed under the prior knowledge of data (e.g., the known label), this kind of method has an alternative name—supervised learning approaches. Figure 1.1.4 presents an example of supervised learning based on prior knowledge—label, where the positive and negative objects are marked by round and cross symbols respectively. The aim of classification is to build up a dividing line to differentiate the positive and negative points from the existing labels. A number of classification algorithms have been well studied in data mining and machine learning domains, the common and well used approaches include Decision Trees, Rule-based Induction, Genetic Algorithms, Neural Networks, Bayesian Networks, Support Vector Machine (SVM), C4.5 and so on. Figure 1.1.5 is a constructed decision tree from the observations of whether it is appropriate to play tennis depending on the weather conditions, such as sunny, rainy, windy, humid conditions and so on. In this example, the classification rules are expressed as a set of If-Then clauses. Apart from decision tree, classifier is another important classification model. Based on the different classification requirement, various classifiers could be trained upon the supervision, e.g., Fig. 1.1.6 demonstrates an example of linear and nonlinear classifier in the above example of debt-income relationship case.

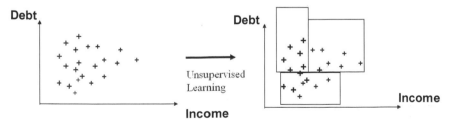

Figure 1.1.3: Example of unsupervised learning

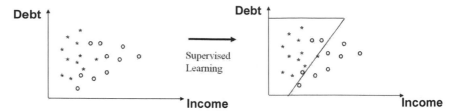

Figure 1.1.4: Example of supervised learning

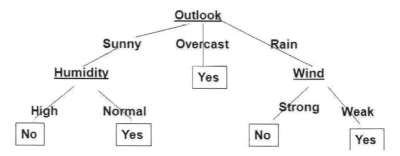

Figure 1.1.5: Example of decision tree

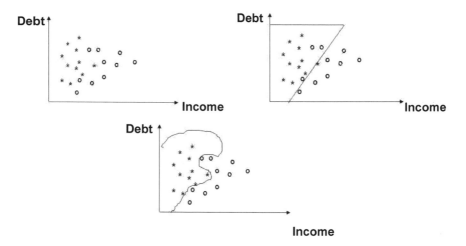

Figure 1.1.6: Linear and nonlinear classification

1.1.3.5 Advanced Data Mining Algorithms

Despite the great success of data mining techniques applied in different areas and settings, there is an increasing demand for developing new data mining algorithms and improving state-of-the-art approaches to handle the more complicated and dynamical problems. In the meantime, with the prevalence and deployment of data mining in real applications, some new research questions and emerging research directions have been raised in response to the advance and breakthrough of theory and technology in data mining. Consequently, applied data mining is becoming an active and fast progressing topic which has opened up a big algorithmic space and developing potential. Here we list some interesting topics, which will be described in subsequent chapters.

1. **High-Dimensional Clustering** In general, data objects to be clustered are described by points in a high-dimensional space, where each dimension corresponds to an attribute/feature. A distance measurement between any two points is used to measure their similarity. The research has shown that the increasing dimensionality results in the loss of contrast in distances between data objects. Thus, clustering algorithms that measure the similarity between data objects based on all attributes/features tend to degrade in high dimensional data spaces. In additional, the widely used distance measurement usually perform effectively only on some particular subsets of attributes, where the data objects are distributed densely. In other words, it is more likely to form dense and reasonable clusters of data objects in a low-dimensional subspace. Recently, several algorithms for discovering data object clusters in subsets of attributes have been proposed, and they can be classified into two categories: *subspace clustering* and *projective clustering* [8].

2. **Multi-Label Classification** In the framework of classification, each object is described as an instance, which is usually a feature vector that characterizes the object from different aspects. Moreover, each instance is associated with one or more labels indicating its categories. Generally speaking, the process of classification consists of two main steps: the first is training a classifier or model on a given set of labeled instances, the second is using the learned classifier to predict the label of unseen instance. However, the instances might be assigned with multiple labels simultaneously, and problems of this type are ubiquitous in many modern applications. Recently, there has been a considerable amount of research concerned with dealing with multi-label problems and many state-of-the-art methods have already been proposed [3]. It has also been applied to lots of practical applications, including text classification, gene function prediction, music emotion analysis, semantic annotation of video, tag recommendation, etc.

3. **Stream data mining** Data stream mining is an important issue because it is the basis for many applications, such as network traffic, web searches, sensor network processing, etc. The purpose of data stream mining is to discover the patterns or structures from the continuous data, which may be used later to infer events that could happen. The special characteristics for stream data is its dynamics that commonly stream data can be read only once. This property limits many traditional strategies for analyzing stream data, because these works always assume that the whole data could be stored in limited storage. In other words, stream data mining could be thought as computation on very large (unlimited large) data.

4. **Recommender Systems** These are important applications because they are essential for many business models. The purpose of recommender systems is to suggest some good items to people based on their preference and historical purchased data. The basic idea of these systems is that if users shared the same interests in the past, they will, with high probability, have similar behaviors in the future. The historical data which reflects users' preferences may consist of explicit ratings, web click log, or tags [6]. It is obviously that personalization plays a critical role in an effective recommendation system [7].

1.2 Organization of the Book

This book is structured into three parts. Part 1: Fundamentals, Part 2: Advanced Data Mining and Part 3: Emerging Applications. In Part 1, we mainly introduce and review the fundamental concepts and mathematical models which are commonly used in data mining. Starting from various data types, we introduce the basic measures and data preprocessing techniques applied in data mining. This part includes five chapters, which will lay down a solid base and prepare the necessary skills and approaches for further understanding the subsequent chapters. Part 2 covers three chapters and addresses the topics of advanced clustering, multi-label classification and stream data mining, which are all hot topics in applied data mining. In addition, we report some recently emerging application directions in applied data mining. Particularly, we will discuss the issues of privacy preserving, recommender systems and social tagging annotation systems, where we will structure the contents in a sequence of theoretical background, state-of-the-art techniques, application cases and future research questions. We also aim to highlight the applied potential of these challenging topics.

1.2.1 Part 1: Fundamentals

1.2.1.1 Chapter 2

Mathematics plays an important role in data mining. As a handbook covering a variety of research topics mentioned in related disciplines, it is necessary to prepare some basic but crucial concepts and backgrounds for readers to easily proceed to the following chapters. This chapter forms an essential and solid base to the whole book.

1.2.1.2 Chapter 3

Data preparation is the beginning of the data mining process. Data mining results are heavily dependent on the data quality prepared before the mining process. This chapter discusses related topics with respect to data preparation, covering attribute selection, data cleaning and integrity, data federation and integration, etc.

1.2.1.3 Chapter 4

Cluster analysis forms the topic of Chapter 4. In this chapter, we classify the proposed clustering algorithms into four categories: traditional clustering algorithm, high-dimensional clustering algorithm, constraint-based clustering algorithm, and consensus clustering algorithm. The traditional data clustering approaches include partitioning methods, hierarchical methods, density-based methods, grid-based methods, and model-based methods. Two different kinds of high-dimensional clustering algorithms are also described. In the constraint-based clustering algorithm subsection, the concept is defined; the algorithms are described and comparison of different algorithms are presented as well. Consensus clustering algorithm is based on the clustering results and is a new way to find robust clustering results.

1.2.1.4 Chapter 5

Chapter 5 describes the methods for data classification, including decision tree induction, Bayesian network classification, rule-based classification, neural network technique of back-propagation, support vector machines, associative classification, k-nearest neighbor classifiers, case-based reasoning, genetic algorithms, rough set theory, and fuzzy set approaches. Issues regarding accuracy and how to choose the best classifier are also discussed.

1.2.1.5 Chapter 6

The original motivation for searching frequent patterns (i.e., association rules) came from the need to analyze supermarket transaction data, which is indeed to discover customer behavior based on the purchased items. Association rules present the fact that how frequently items are bought together. For example, an association rule "beer-diaper (70%)" indicates that 70% of the customers who bought beer also bought diapers. Such rules can be used to make predictions and recommendations for customers and design then store layout. Stemming from the basic itemset data, rule discovery on more general and complex data (i.e., sequence, tree, graph) has been thoroughly explored for the past decade. In this chapter, we introduce the basic techniques of frequent pattern mining on different types of data, i.e., itemset, sequence, tree, and graph.

1.2.2 Part 2: Advanced Data Mining

1.2.2.1 Chapter 7

This chapter reports the latest research progress in clustering analysis from three different aspects: (1) improve the clustering result quality of heuristic clustering algorithm by using Space Smoothing Search methods; (2) use approximate backbone to capture the common optimal information of a given data set, and then use the approximate backbone to improve the clustering result quality of heuristic clustering algorithm; (3) design a local significant unit (LSU) structure to capture the data distribution in high-dimensional space to improve the clustering result quality based on kernel estimation and spatial statistical theory.

1.2.2.2 Chapter 8

Recently, there has been a considerable amount of research dealing with multi-label problems and many state-of-the-art methods have already been proposed. It has also been applied to lots of practical applications. In this chapter, a comprehensive and systematic study of multi-label classification is carried out in order to give a clear description of what multi-label classification is, and what are the basic and representative methods, and what are the future open research questions.

1.2.2.3 Chapter 9

Data stream mining is the process of discovering structures or rules from rapid continuous data, which can commonly be read only once with limited storage capabilities. The issue is important because it is the basis of many real applications, such as sensor network data, web queries, network traffic, etc. The purpose of the study on data stream mining is to make appropriate predictions, by exploring the historical stream data. In this chapter, we present the main techniques to tackle the challenge.

1.2.3 Part 3: Emerging Applications

1.2.3.1 Chapter 10

Privacy-preserving data mining is an important issue because there is an increasing requirement of storing personal data for users. The issue has been thoroughly studied in several areas such as the database community, the cryptography community, and the statistical disclosure control community. In this chapter, we will discuss the basic concepts and main strategies of privacy-preserving data mining.

1.2.3.2 Chapter 11

Recommender systems present people with interesting items based on information from other people. The basic idea of these systems is that if users shared the same interests in the past, they will also have similar behaviors in the future. The information that other people provide may come from explicit ratings, tags, or reviews. Specially, the recommendations may be personalized to the preferences of different users. In this chapter, we introduce the basic concepts and strategies for recommender systems.

1.2.3.3 Chapter 12

With the popularity of social web technologies social tagging systems have become an important application and service. The social web data produced by the collaborative practice of mass provides a new arena in data mining research. One emerging research trend in social web mining is to make use of the tagging behavior in social annotation systems for presenting the most demanded information to users—i.e., personalized recommendations. In this chapter, we aim at bridging the gap between social tagging systems and recommender systems. After introducing the basic concepts in social collaborative annotation systems and reviewing the advances in recommender systems, we address the research issues of social tagging recommender systems.

1.3 The Audience of the Book

This book not only combines the fundamental concepts, models and algorithms in the data mining domain together to serve as a referential handbook to researchers and practitioners from as diverse backgrounds as Computer Science, Machine Learning, Information Systems, Artificial Intelligence, Statistics, Operational Science, Business Intelligence as well as Social Science disciplines but also provides a compilation and summarization for disseminating and reviewing the recently emerging advances in a variety of data mining application arenas, such as Advanced Data Mining, Analytics, Internet Computing, Recommender Systems, Information Retrieval as well as Social Computing and Applied Informatics from the perspective of developmental practice for emerging researches and real applications. This book will also be useful as a text book for postgraduate students and senior undergraduate students in related areas.

The salient features of this book is that it:

- Systematically presents and discusses the mathematical background and representative algorithms for Data Mining, Information Retrieval and Internet Computing.
- Thoroughly reviews the related studies and outcomes conducted on the addressed topics.
- Substantially demonstrates various important applications in the areas of classical Data Mining, Advanced Data Mining and emerging research topics such as Privacy Preserving, Stream Data Mining, Recommender Systems, Social Computing etc.
- Heuristically outlines the open research questions of interdisciplinary research topics, and identifies several future research directions that readers may be interested in.

References

[1] R. Agrawal, R. Srikant et al. Fast algorithms for mining association rules. *In: Proc. 20th Int. Conf. Very Large Data Bases, VLDB*, Vol. 1215, pp. 487–99, 1994.
[2] M. Anderberg. Cluster analysis for applications. Technical report, DTIC Document, 1973.
[3] B. Fu, Z. Wang, R. Pan, G. Xu and P. Dolog. Learning tree structure of label dependency for multi-label learning. *In: PAKDD (1)*, pp. 159–70, 2012.
[4] J. Han, H. Cheng, D. Xin and X. Yan. Frequent pattern mining: current status and future directions. *Data Mining and Knowledge Discovery*, 15(1): 55–86, 2007.
[5] J. Han and M. Kamber. *Data Mining: Concepts and Techniques*. Morgan Kaufmann, 2006.
[6] G. Xu, Y. Gu, P. Dolog, Y. Zhang and M. Kitsuregawa. Semrec: a semantic enhancement framework for tag based recommendation. *In: Proceedings of the Twenty-fifth AAAI Conference on Artificial Intelligence (AAAI-11)*, 2011.
[7] G. Xu, Y. Zhang and L. Li. *Web Mining and Social Networking: Techniques and Applications*, Vol. 6. Springer, 2010.
[8] Y. Zong, G. Xu, P. Jin, X. Yi, E. Chen and Z. Wu. A projective clustering algorithm based on significant local dense areas. *In: Neural Networks (IJCNN), The 2012 International Joint Conference on*, pp. 1–8. IEEE, 2012.

CHAPTER 2

Mathematical Foundations

Data mining is a data analysis process involving in data itself, operators and various numeric metrics. Before we go deeply into the algorithm and technique part, we first summarize and present some relevant basic but important expressions and concepts from mathematical books and open available sources (e.g., Wikipedia).

2.1 Organization of Data

As mentioned earlier, data sets come in different forms [1]: these forms are known as schemas. The simplest form of data is a set of vector measurements on objects $o(1), \cdots, o(n)$. For each object we have measurements of p variables $X_1, \cdots, X_p$. Thus, the data can be viewed as a matrix with n rows and p columns. We refer to this standard form of data as a *data matrix*, or simply *standard data*. We can also refer to data set as a *table*.

Often there are several types of objects we wish to analyze. For example, in a payroll database, we might have data both of employees, with variables of name, department-name, age and salary, and about departments with variables such as department-name, budget and manager. These data matrices are connected to each other. Data sets consisting of several such matrices or tables are called *multi-relational data*.

But some data sets do not fit well into the matrix or table form. A typical example is a time series, which can use only a related ordered data type named event-sequence. In some applications, there are more complex schemas, such as graph-based model, hierarchical structure, etc.

To summarize, in any data mining application it is crucial to be aware of the schema of the data. Without such an awareness, it is easy to miss important patterns in the data, or perhaps worse, to rediscover patterns that are part of the fundamental design of the data. In addition, we must be particularly careful about data schemas.

2.1.1 Boolean Model

There is no doubt that the Boolean model is one of the most useful random set models in mathematical morphology, stochastic geometry and spatial statistics. It is defined as the union of a family of independent random compact subsets (denoted in short as "objects") located at the points of a locally finite Poisson process. It is stationary if the objects are identically distributed (up to their location) and the Poisson process is homogeneous, otherwise it is non-stationary. Because the definition of set is very intuitive, the Boolean model provides an uncomplicated framework for information retrieval system users. Unfortunately, the Boolean model has some drawbacks. First, the search strategy is based on binary criteria, the lack of the concept of document classification is well known, so the search function is limited. Second, Boolean expressions have precise semantics, but it is often difficult to convert the user's information to Boolean expressions. In fact, most users find it is not so easily to converted to a Boolean query information they need. To get rid of these defects, Boolean model is still the main model for document database system. The major advantage of the Boolean model has a clear and simple form, but the major drawback is that complete match will lead to a result of too much or too little of the document being returned. As we all know, the weight of the index terms fundamentally improves the function of the retrieval system, resulting in the generation of the vector model.

2.1.2 Vector Space Model

Vector space model is an algebraic model for representing text documents (and any object in general) as vectors of identifiers, such as, for example, index terms. It is used in information filtering, information retrieval, indexing and relevancy rankings. In vector space model, documents and queries are represented as vectors.

Each dimension corresponds to a separate term. The definition of term depends on the application. Typically, terms are single words, keywords, or longer phrases. If words are chosen to be the terms, the dimensionality of the vector is the number of words in the vocabulary (the number of distinct words occurring in the corpus).

If a term occurs in the document, its value in the vector is non-zero. Several different ways of computing these values, also known as (term) weights, have been developed. One of the best known schemes is *tf-idf* weighting, and the model is known as *term frequency-inverse document frequency model*. Unlike the term count model, *tf-idf* incorporates local and global information. The weighted vector for document d is $v_d = (w_{1,d}, w_{2,d}, \cdots, w_{N,d})^T$, where term weight is defined as:

$$w_{i,d} = tf_{i,d} * \log\left(\frac{D}{df_{i,d}}\right) \tag{2.1.1}$$

where $tf_{i,d}$ is the term frequency (term counts) or number of times a term i occurs in a document. This accounts for local information; $df_{i,d}$ = document frequency or number of documents containing term i; and D= number of documents in a database.

As a basic model, the term vector scheme discussed above has several limitations. First, it is very calculation intensive. From the computational standpoint it is very slow, requiring a lot of processing time. Second, each time we add a new term into the term space we need to recalculate all the vectors. For example, computing the length of the query vector requires access to every document term and not just the terms specified in the query. Other limitations include long documents, false negative matches, semantic content, etc. Therefore, this model can have a lot of improvement space.

2.1.3 Graph Model

Graph is a combination of nodes and edges. The nodes represent different objects while edges are the inter-connection among them. In mathematics, a graph is a pair $G = (V,E)$ of sets such that $E \subseteq [V]^2$. The elements of V are the nodes of the graph G, the elements of E are its edges. Figure 2.1.1 depicts an example of a Graph model.

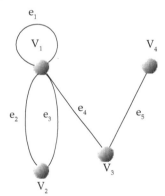

Figure 2.1.1: Example of a Graph model

The most typical graph in real world is the global internet, in which computers, routers and switches are the nodes while network wires or wireless connections are the edges. Similar data sets are easily depicted in the form of graph models, since it is one of the most convenient and illustrative mathematical models to describe the real world phenomenons.

An important concept in the graph model is the adjacent matrix, usually noted as $A = (A_{ij})$

$$A_{ij} = \begin{cases} 1, & i \sim j \\ 0, & \text{otherwise} \end{cases} \qquad (2.1.2)$$

Here, the sign $i \sim j$ means there is an edge between the two nodes. The adjacent matrix contains the structural information of the whole network, moreover, it has a matrix format fitting in both simple and complex mathematical analysis. For a general case extended, we have A defined as

$$A_{ij} = \begin{cases} w_{ij}, & i \sim j \\ 0, & \text{otherwise} \end{cases} \qquad (2.1.3)$$

in which w_{ij} is the weight parameter of the edge between i and j. The basic point of this generalization is the quantification on the strength of the edges in different positions.

Another important matrix involved is the Laplacian matrix $\mathcal{L} = D - A$. Here, $D = \text{Diag}(d_1, \cdots, d_n)$ is the *diagonal degree matrix* where $d_i = \Sigma_{j=1}^{n} A_{ij}$ is the degree of the node i. Scientists use this matrix to explore the structure, like communities or synchronization behaviors, of graphs with appropriate mathematical tools.

Notice that if the graph is undirected, we have $A_{ij} = A_{ji}$, or on the other side, two nodes share different influence from each other, which form a directed graph.

Among those specific graph models, trees and forests are the most studied and applied. An acyclic graph, one not containing any cycles, is called a forest. A connected forest is called a tree. (Thus, a forest is a graph whose components are trees.) The vertices of degree 1 in a tree are its leaves.

Another important model is called Bipartite graphs. The vertices in a Bipartite graph can be divided into two disjoint sets U and V such that every edge connects a vertex in U to one in V; that is, U and V are independent sets. Equivalently, a bipartite graph is a graph that does not contain any odd-length cycles. Figure 2.1.2 shows an example of a Bipartite graph:

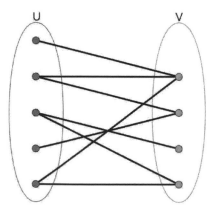

Figure 2.1.2: Example of Bipartite graph

The two sets U and V may be thought of as of two colors: if one colors all nodes in U blue, and all nodes in V green, each edge has endpoints of differing colors, as is required in the graph coloring problem. In contrast, such a coloring is impossible in the case of a non-bipartite graph, such as a triangle: after one node is colored blue and another green, the third vertex of the triangle is connected to vertices of both colors, preventing it from being assigned either color. One often writes $G = (U, V, E)$ to denote a Bipartite graph whose partition has the parts U and V. If $|U| = |V|$, that is, if the two subsets have equal cardinality, then G is called a Balanced Bipartite graph.

Also, scientists have established the Vicsek model to describe swarm behavior. A swarm is modeled in this graph by a collection of particles that move with a constant speed but respond to a random perturbation by adopting at each time increment the average direction of motion of the other particles in their local neighborhood. Vicsek model predicts that swarming animals share certain properties at the group level, regardless of the type of animals in the swarm. Swarming systems give rise to emergent behaviors which occur at many different scales, some of which are turning out to be both universal and robust, as well an important data representation.

PageRank [2] is a link analysis algorithm, used by the Google Internet search engine, that assigns a numerical weight to each element of a hyperlinked set of documents, such as the World Wide Web, with the purpose of "measuring" its relative importance within the set. The algorithm may be applied to any collection of entities with reciprocal quotations and references. A PageRank results from a mathematical algorithm based on the web-graph, created by all World Wide Web pages as nodes and hyperlinks as edges, taking into consideration authority hubs such as cnn.com or usa.gov. The rank value indicates an importance of a particular page. A hyperlink to a page counts as a vote of support. The PageRank of a page is defined recursively and depends on the number and PageRank metric of all pages that link to it ("incoming links"). A page that is linked by many pages with high PageRank receives a high rank itself. If there are no links to a web page there is no support for that page. The following Fig. 2.1.3 shows an example of a PageRank:

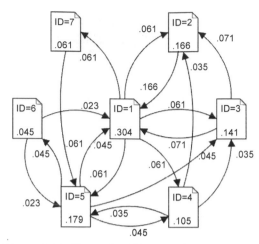

Figure 2.1.3: Example of PageRank execution

2.1.4 Other Data Structures

Besides relational data schemas, there are many other kinds of data that have versatile forms and structures and rather different semantic meanings. Such kinds of data can be seen in many applications: time-related or sequence data (e.g., historical records, stock exchange data, and time-series and biological sequence data), data streams (e.g., video surveillance and sensor data, which are continuously transmitted), spatial data (e.g., maps), engineering design data (e.g., the design of buildings, system components, or inter-rated circuits), hypertext and multimedia data (including text, image, video, and audio data). These applications bring about new challenges, like how to handle data carrying special structures (e.g., sequences, trees, graphs, and networks) and specific semantics (such as ordering, image, audio and video contents, and connectivity), and how to mine patterns that carry rich structures and semantics.

It is important to keep in mind that, in many applications, multiple types of data are present. For example, in informatics, genomic sequences, biological networks, and 3-D spatial structures of genomes may co-exist for certain biological objects. Mining multiple data sources of complex data often leads to fruitful findings due to the mutual enhancement and consolidation of such multiple sources. On the other hand, it is also challenging because of the difficulties in data cleaning and data integration, as well as the complex interactions among the multiple sources of such data. While such data require sophisticated facilities for efficient storage, retrieval, and updating, they also provide fertile ground and raise challenging research and implementation issues for data mining. Data mining on such data is an advanced topic.

2.2 Data Distribution

2.2.1 Univariate Distribution

In probability and statistics, a univariate distribution [3] is a probability distribution of only one random variable. This is in contrast to a multivariate distribution, the probability distribution of a random vector.

A random variable or stochastic variable is a variable whose value is subject to variations due to chance (i.e., randomness, in a mathematical sense). As opposed to other mathematical variables, a random variable conceptually does not have a single, fixed value (even if unknown); rather, it can take on a set of possible different values, each with an associated probability. The interpretation of a random variable depends on the interpretation of probability:

- *The objectivist viewpoint*: As the outcome of an experiment or event where randomness is involved (e.g., the result of rolling a dice, which is a number between 1 and 6, all with equal probability; or the sum of the results of rolling two dices, which is a number between 2 and 12, with some numbers more likely than others).
- *The subjectivist viewpoint*: The formal encoding of one's beliefs about the various potential values of a quantity that is not known with certainty (e.g., a particular person's belief about the net worth of someone like Bill Gates after Internet research on the subject, which might have possible values ranging between 50 billion and 100 billion, with values near the center more likely).
- Random variables can be classified as either *discrete* (i.e., it may assume any of a specified list of exact values) or *continuous* (i.e., it may assume any numerical value in an interval or collection of intervals). The mathematical function describing the possible values of a random variable and their associated probabilities is known as a probability distribution. The realizations of a random variable, i.e., the results of randomly choosing values according to the variable's probability distribution are called *random variates*.

A random variable's possible values might represent the possible outcomes of a yet-to-be-performed experiment or an event that has not happened yet, or the potential values of a past experiment or event whose already-existing value is uncertain (e.g., as a result of incomplete information or imprecise measurements). They may also conceptually represent either the results of an "objectively" random process (e.g., rolling a dice), or the "subjective" randomness that results from incomplete knowledge of a quantity. The meaning of the probabilities assigned to the potential values of a random variable is not part of probability theory itself, but instead related to philosophical arguments over the interpretation of probability. The mathematics works the same regardless of the particular interpretation in use.

The basic concept of "random variable" in statistics is real-valued. However, one can consider arbitrary types such as boolean values, complex numbers, vectors, matrices, sequences, trees, sets, shapes, manifolds, functions, and processes. The term "random element" is used to encompass all such related concepts. A related concept is the stochastic process, a set of indexed random variables (typically indexed by time or space). This more general concept is particularly useful in fields such as computer science and natural language processing where many of the basic elements of analysis are non-numerical. These general random variables are typically parameterized as sets of real-valued random variables often more specifically as random vectors.

2.2.2 Multivariate Distribution

In probability theory and statistics, the multivariate normal distribution or other multivariate distribution model [4], such as multivariate complex Gaussian distribution, is a generalization of the one-dimensional (univariate) normal distribution to higher dimensions. One possible definition is that a random vector is said to be p-variate normally distributed if every linear combination of its p components has a univariate normal distribution. However, its importance derives mainly from the multivariate central limit theorem. The multivariate normal distribution is often used to describe, at least approximately, any set of (possibly) correlated real-valued random variables each of which clusters around a mean value.

The multivariate normal distribution is undoubtedly one of the most well-known and useful distributions in statistics, playing a predominant role in many areas of applications. In multivariate analysis, for example, most of the existing inference procedures for analyzing vector-valued data have been developed under the assumption of normality. In linear model problems, such as the analysis of variance and regression analysis, the error vector is often assumed to be normally distributed so that statistical analysis can be performed using distributions derived from the normal distribution. In addition to appearing in these areas, the multivariate normal distribution also appears in multiple comparisons, in the studies of dependence of random variables, and in many other related areas.

There are, of course, many reasons for the predominance of the multivariate normal distribution in statistics. These result from some of its most desirable properties as listed below:

1. It represents a natural extension of the univariate normal distribution and provides a suitable model for many real-life problems concerning vector-valued data.
2. Even if in an experiment, the original data cannot be fitted satisfactorily with a multivariate normal distribution (as is the case when the measurements are discrete random vectors), by the central limit theorem, the distribution of the sample mean vector is asymptotically normal. Thus the multivariate normal distribution can be used for approximating the distribution of the same mean vector in the large sample case.
3. The density function of a multivariate normal distribution is uniquely determined by the mean vector and the covariance matrix of the random variable.
4. Zero correlations imply independence; that is, if all the correlation coefficients between two sets of components of a multivariate normal variable are zero, then the two sets of components are independent.
5. The family of multivariate normal distributions is closed under linear transformations and linear combinations. In other words, the distributions of linear transformations or linear combinations of multivariate normal variables are again multivariate normal.
6. The marginal distribution of any subset of components of a multivariate normal variable is also multivariate normal.
7. The conditional distribution in a multivariate normal distribution is multivariate normal. Furthermore, the conditional mean vector is a linear function and the conditional covariance matrix depends only on the covariance matrix of the joint distribution. This property yields simple and useful results in regression analysis and correlation analysis.
8. For the bivariate normal distribution, positive and negative dependence properties of the components of a random vector are completely determined by the sign and the size of the correlation coefficient. Similar results also exist for the multivariate normal distribution. Thus it is often chosen as an ideal model for studying the dependence of random variables.

2.3 Distance Measures

We now take a short detour to study the general notion of distance measures [5].

2.3.1 Jaccard distance

The Jaccard similarity is a measure of how close sets are, although it is not really a distance measure. That is, the closer the sets are, the higher the Jaccard similarity, which is

$$J(A,B) = \frac{|A \cap B|}{|A \cap B|}. \qquad (2.3.1)$$

Rather, 1 minus the Jaccard similarity is a distance measure, as we shall see; it is called the Jaccard distance:

$$J_\delta(A, B) = 1 - J(A, B) = \frac{|A \cup B| - |A \cap B|}{|A \cup B|}. \qquad (2.3.2)$$

However, Jaccard distance is not the only measure of closeness that makes sense. We shall examine in this section some other distance measures that have applications.

2.3.2 Euclidean Distance

The most familiar distance measure is the one we normally think of as "distance". An n-dimensional Euclidean space is one where points are vectors of n real numbers. The conventional distance measure in this space, which we shall refer to as the L2-norm, is defined:

$$d\left([x_1, x_2, \cdots, x_n], [y_1, y_2, \cdots, y_n]\right) = \sqrt{\sum_{i=1}^{n}(x_i - y_i)^2}. \qquad (2.3.3)$$

That is, we square the distance in each dimension, sum the squares, and take the positive square root.

It is easy to verify that the first three requirements for a distance measure are satisfied. The Euclidean distance between two points cannot be negative, because the positive square root is intended. Since all squares of real numbers are nonnegative, any i such that $x_i = y_i$ forces the distance to be strictly positive. On the other hand, if $x_i = y_i$ for all i, then the distance is clearly 0. Symmetry follows because $(x_i - y_i)^2 = (y_i - x_i)^2$. The triangle inequality requires a good deal of algebra to verify. However, it is well understood to be a property of Euclidean space: the sum of the lengths of any two sides of a triangle is no less than the length of the third side.

There are other distance measures that have been used for Euclidean spaces. For any constant r, we can define the Lr-norm to be the distance measure defined by:

$$d\left([x_1, x_2, \cdots, x_n], [y_1, y_2, \cdots, y_n]\right) = \left(\sqrt{\sum_{i=1}^{n} |x_i - y_i|^r}\right)^{1/r}. \quad (2.3.4)$$

The case $r = 2$ is the usual L2-norm just mentioned. Another common distance measure is the L1-norm, or Manhattan distance. There, the distance between two points is the sum of the magnitudes of the differences in each dimension. It is called "Manhattan distance" because it is the distance one would have to travel between points if one were constrained to travel along grid lines, as on the streets of a city such as Manhattan.

Another interesting distance measure is the L∞ -norm, which is the limit as r approaches infinity of the Lr-norm. As r gets larger, only the dimension with the largest difference matters, so formally, the L∞ -norm is defined as the maximum of $|x_i - y_i|$ over all dimensions i.

2.3.3 Minkowski Distance

The Minkowski distance is a metric on Euclidean space which can be considered as a generalization of the Euclidean distance. The Minkowski distance of order p between two points

$$P = (x_1, x_2, \cdots, x_n) \text{ and } Q = (y_1, y_2, \cdots, y_n) \in R^n \quad (2.3.5)$$

is defined as:

$$\left(\sum_{i=1}^{n} |x_i - y_i|^p\right)^{1/p} \quad (2.3.6)$$

The Minkowski distance is a metric as a result of the Minkowski inequality. Minkowski distance is typically used with p being 1 or 2. The latter is the Euclidean distance, while the former is sometimes known as the Manhattan distance. In the limiting case of p reaching infinity we obtain the Chebyshev distance:

$$\lim_{p \to \infty} \left(\sum_{i=1}^{n} |x_i - y_i|^p\right)^{1/p} = \max_{i=1}^{n} |x_i - y_i| \quad (2.3.7)$$

Similarly, when p reaches negative infinity we have

$$\lim_{p \to -\infty} \left(\sum_{i=1}^{n} |x_i - y_i|^p\right)^{1/p} = \min_{i=1}^{n} |x_i - y_i| \quad (2.3.8)$$

The Minkowski distance is often used when variables are measured on ratio scales with an absolute zero value. Variables with a wider range can overpower the result. Even a few outliers with high values bias the result and disregard the alikeness given by a couple of variables with a lower upper bound.

2.3.4 Chebyshev Distance

In mathematics, Chebyshev distance is a metric defined on a vector space where the distance between two vectors is the greatest of their differences along any coordinate dimension. It is also known as chessboard distance, since in the game of chess the minimum number of moves needed by the king to go from one square on a Chessboard to another equals the Chebyshev distance between the centers of the squares. The Chebyshev distance between two vectors or points p and q, with standard coordinates p_i and q_i, respectively, is

$$D_{\text{Chebyshev}}(p, q) = \max_i | p_i - q_i |$$ (2.3.9)

This equals the limit of the L_p metrics. In one dimension, all L_p metrics are equal—they are just the absolute value of the difference:

$$\lim_{k \to \infty} (\sum_{i=1}^{n} | p_i - q_i |^k)^{1/k}.$$ (2.3.10)

Mathematically, the Chebyshev distance is a metric induced by the supremum norm or uniform norm. It is an example of an injective metric. In two dimensions, i.e., plane geometry, if the points p and q have Cartesian coordinates (x_1, y_1) and (x_2, y_2), their Chebyshev distance is

$$D_{\text{Chess}} = \max(| x_2 - x_1 |, | y_2 - y_1 |).$$ (2.3.11)

In fact, Manhattan distance, Euclidean distance above and Chebyshev distance are Minkowski distance in special conditions.

2.3.5 Mahalanobis Distance

In statistics, Mahalanobis distance is another distance measure. It is based on correlations between variables by which different patterns can be identified and analyzed. It gauges similarity of an unknown sample set to a known one. It differs from Euclidean distance in that it takes into account the correlations of the data set and is scale-invariant. In other words, it is a multivariate effect size. Formally, the Mahalanobis distance of a multivariate vector $x = (x_1, x_2, x_3, \cdots, x_N)^T$ from a group of values with mean $\mu = (\mu_1, \mu_2, \mu_3, \cdots, \mu_N)^T$ and covariance matrix S is defined as:

$$D_M(x) = \sqrt{(x - \mu)^T S^{-1} (x - \mu)}.$$ (2.3.12)

Mahalanobis distance can also be defined as a dissimilarity measure between two random vectors $\overrightarrow{x}$ and $\overrightarrow{y}$ of the same distribution with the covariance matrix S:

$$d(\vec{x}, \vec{y}) = \sqrt{(\vec{x} - \vec{y})^T S^{-1} (\vec{x} - \vec{y})}. \qquad (2.3.13)$$

If the covariance matrix is the identity matrix, the Mahalanobis distance reduces to the Euclidean distance. If the covariance matrix is diagonal, then the resulting distance measure is called the normalized Euclidean distance:

$$d(\vec{x}, \vec{y}) = \sqrt{\sum_{i=1}^{N} \frac{(x_i - y_i)^2}{s_i^2}}. \qquad (2.3.14)$$

where s_i is the standard deviation of the x_i and y_i over the sample set. Mahalanobis' discovery was prompted by the problem of identifying the similarities of skulls based on measurements. And now, it is widely used in cluster analysis and classification techniques.

2.4 Similarity Measures

2.4.1 Cosine Similarity

In some applications, the classic vector space model is used generally, such as Relevance rankings of documents in a keyword search. It can be calculated, using the assumptions of document similarities theory, by comparing the deviation of angles between each document vector and the original query vector where the query is represented as same kind of vector as the documents.

An important problem that arises when we search for similar items of any kind is that there may be far too many pairs of items to test each pair for their degree of similarity, even if computing the similarity of any one pair can be made very easy. Finally, we explore notions of "similarity" that are not expressible as inter-section of sets. This study leads us to consider the theory of distance measures in arbitrary spaces. Cosine similarity is often used to compare documents in text mining.

In addition, it is used to measure cohesion within clusters in the field of data mining. The cosine distance makes sense in spaces that have dimensions, including Euclidean spaces and discrete versions of Euclidean spaces, such as spaces where points are vectors with integer components or boolean (0 or 1) components. In such a space, points may be thought of as directions. We do not distinguish between a vector and a multiple of that vector. Then the cosine distance between two points is the angle that the vectors to those points make. This angle will be in the range of 0° to 180°, regardless of how many dimensions the space has.

We can calculate the cosine distance by first computing the cosine of the angle, and then applying the arc-cosine function to translate to an angle in the 0–180° range. Given two vectors x and y, the cosine of the angle between them is the dot product of x and y divided by the L2-norms of x and y (i.e., their Euclidean distances from the origin). Recall that the dot product of vectors $\vec{x} = [x_1, x_2, \cdots, x_n]$ and $\vec{y} = [y_1, y_2, \cdots, y_n]$ is $\sum_{i=1}^{n} x_i * y_i$, the cosine similarity is defined as:

$$CosSim\ (\vec{x}, \vec{y}) = \frac{\vec{x} \cdot \vec{y}}{\| \vec{x} \| * \| \vec{y} \|} \tag{2.4.1}$$

We must show that the cosine similarity is indeed a distance measure. We have defined that the angle of two vector is in the range of 0 to 180, no negative similarity value is possible. Two vectors have an angle of zero if and only if they are along the same direction but with possible different length magnitude. Symmetry is obvious: the angle between x and y is the same as the angle between y and x. The triangle inequality is best argued by physical reasoning.

One way to rotate from x to y is to rotate to z and thence to y. The sum of those two rotations cannot be less than the rotation directly from x to y.

2.4.2 Adjusted Cosine Similarity

Although the prejudices of individuals can be certainly amended by Cosine similarity, but only to distinguish the individual differences between the different dimensional cannot measure the value of each dimension, it would lead to such a situation, for example, the content ratings by 5 stars, two user X and Y, on the two resources ratings are respectively (1, 2) and (4, 5), using the results of the cosine similarity is 0.98, both are very similar. But with the score of X, it seems X don't like these two resources, and Y. The reason for this situation is that likes it more the distance metric is a measure of space between each points' absolute distance with each location coordinates directly; and the cosine similarity measure relies on space vector angle and is reflected in the direction of the difference, not location. So the adjust cosine similarity appeared. All dimension values are subtracted from an average value, such as X and Y scoring average is 3, so after adjustment for (-2, -1) and (1,2), then the cosine similarity calculation, -0.8, similarity is negative and the difference is not small, but clearly more in line with the reality. Based on the above exposition, computing similarity using basic cosine measure in item-based case has one important drawback—the difference in rating scale between different users are not taken into account. The adjusted cosine similarity offsets this drawback by subtracting the corresponding user average from each co-rated pair. Formally, the similarity between items i and j using this scheme is given by

$$\text{sim}(i, j) = \frac{\sum_{u \in U} (R_{u,i} - \overline{R}_u)\,(R_{u,j} - \overline{R}_u)}{\sqrt{\sum_{u \in U} (R_{u,i} - \overline{R}_u)^2}\,\sqrt{\sum_{u \in U} (R_{u,j} - \overline{R}_u)^2}}. \tag{2.4.2}$$

Here $\overline{R}_\mu$ is the average of the u-th user's ratings.

2.4.3 Kullback-Leibler Divergence

In probability theory and information theory, the Kullback-Leibler divergence is a nonsymmetric measure of the difference between two probability distributions P and Q. KL measures the expected number of extra bits required to code samples from P when using a code based on Q, rather than using a code based on P. Typically P represents the "true" distribution of data, observations, or a precisely calculated theoretical distribution. The measure Q typically represents a theory, model, description, or approximation of P.

Although it is often intuited as a metric or distance, the KL divergence is not a true metric—for example, it is not symmetric: the KL from P to Q is generally not the same as the KL from Q to P. However, its infinitesimal form, specifically its Hessian, is a metric tensor: it is the Fisher information metric.

For probability distributions P and Q of a discrete random variable their KL divergence is defined to be

$$D_{KL}(P \| Q) = \sum_i P(i) \ln \frac{P(i)}{Q(i)}. \tag{2.4.3}$$

In words, it is the average of the logarithmic difference between the probabilities P and Q, where the average is taken using the probabilities P. The KL divergence is only defined if P and Q both sum up to 1 and if $Q(i) > 0$ for any i is such that $P(i) > 0$. If the quantity $0 \ln 0$ appears in the formula, it is interpreted as zero. For distributions P and Q of a continuous random variable, KL divergence is defined to be the integral:

$$D_{KL}(P \| Q) = \int_{-\infty}^{\infty} p(x) \ln \frac{p(x)}{q(x)} dx, \tag{2.4.4}$$

where p and q denote the densities of P and Q. More generally, if P and Q are probability measures over a set X, and Q is absolutely continuous with respect to P, then the Kullback-Leibler divergence from P to Q is defined as

$$D_{KL}(P \| Q) = \int_X \ln \frac{dQ}{dP} dP, \tag{2.4.5}$$

where $\dfrac{dQ}{dp}$ is the Radon-Nikodym derivative of Q with respect to P, and provided the expression on the right-hand side exists. Likewise, if P is absolutely continuous with respect to Q, then

$$D_{KL}(P \parallel Q) = \int_X \ln \frac{dP}{dQ} dP = \int_X \frac{dP}{dQ} \ln \frac{dP}{dQ} dQ. \qquad (2.4.6)$$

which we recognize as the entropy of P relative to Q. Continuing in this case, if μ is any measure on X for which $p = \dfrac{dP}{d\mu}$ and $q = \dfrac{dQ}{d\mu}$ exist, then the Kullback-Leibler divergence from P to Q is given as

$$D_{KL}(P \parallel Q) = \int_X p \ln \frac{p}{q} d\mu. \qquad (2.4.7)$$

The logarithms in these formulae are taken to base 2 if information is measured in units of bits, or to base e if information is measured in nats. Most formulas involving the KL divergence hold irrespective of log base. For probability distributions P and Q of a discrete random variable, their KL divergence is defined as

$$D_{KL}(P \parallel Q) = \sum_i p(i) \ln \frac{P(i)}{Q(i)}. \qquad (2.4.8)$$

In other words, it is the average of the logarithmic difference between the probabilities P and Q, where the average is taken using the probabilities P. The KL divergence is only defined if P and Q both sum up to 1 and if $Q(i) > 0$ for any i is such that $P(i) > 0$. If the quantity $0 \ln 0$ appears in the formula, it is interpreted as zero. For distributions P and Q of a continuous random variable, KL divergence is defined to be the integral:

$$D_{KL}(P \parallel Q) = \int_{-\infty}^{\infty} p(x) \ln \frac{p(x)}{q(x)} dx, \qquad (2.4.9)$$

where p and q denote the densities of P and Q. More generally, if P and Q are probability measures over a set X, and Q is absolutely continuous with respect to P, then the Kullback-Leibler divergence from P to Q is defined as

$$D_{KL}(P \parallel Q) = -\int_X \ln \frac{dQ}{dP} dP, \qquad (2.4.10)$$

where $\dfrac{dQ}{dP}$ is the Radon-Nikodym derivative of Q with respect to P, and provided the expression on the right-hand side exists. Likewise, if P is absolutely continuous with respect to Q, then

$$D_{KL}(P \parallel Q) = \int_X \ln \frac{dP}{dQ} dP \int_X \frac{dP}{dQ} \ln \frac{dP}{dQ} dQ, \qquad (2.4.11)$$

which we recognize as the entropy of P relative to Q. Continuing in this case, if μ is any measure on X for which $p = \frac{dP}{d\mu}$ and $q = \frac{dQ}{d\mu}$ exist, then the Kullback-Leibler divergence from P to Q is given as

$$D_{KL}(P \parallel Q) = \int_X p \ln \frac{p}{q} d\mu, \qquad (2.4.12)$$

The logarithms in these formulae are taken to base 2 if information is measured in units of bits, or to base e if information is measured in nats. Most formulas involving the KL divergence hold irrespective of log base. The Kullback-Leibler divergence is a widely used tool in statistics and pattern recognition. In Bayesian statistics the KL divergence can be used as a measure of the information gain in moving from a prior distribution to a posterior distribution. And the KL divergence between two Gaussian Mixture Models (GMMs) is frequently needed in the fields of speech and image recognition.

2.4.4 Model-based Measures

Distance or similarity functions play a central role in all clustering algorithms. Numerous distance functions have been reported in the literature and used in applications. Different distance functions are also used for different types of attributes (also called variables). A most commonly used distance functions for numeric attributes is Manhattan (city block) distance. This distance measures in special cases of a more general distance function is called the Minkowski distance. But the above distance measures are only appropriate for numeric attributes. For binary and nominal attributes (also called unordered categorical attributes), we need different functions. Thus, an algorithm might be required to test the similarity functions for their appropriation on different specific models or attributes. Below is a popular method to establish the accuracy level of similarity functions.

Consider a graph $G(V, E)$ with the same definition mentioned above. The multiple links and self-connections are not allowed. For each pair of nodes, $x, y \in V$, we assign a score, s_{xy}, according to a given similarity measure. Higher score means higher similarity between x and y, and vice versa. Suppose G is undirected, the score is also supposed to be symmetry as the adjacent matrix, say $s_{xy} = s_{yx}$. All the nonexistent links are sorted in a descending order according to their scores, and the links at the top are most likely to exist. To test the algorithm's accuracy, the observed links, E, is randomly divided into two parts: the training set, E^T, is treated as known information, while the probe set, E^P, is used for testing and no information therein is allowed to be used for prediction. Clearly, $E = E^T \cup E^P$ and $E^T \cap E^P = \varnothing$. We can choose different portion rate of these two sets for the test. To quantify the prediction accuracy, we use a standard metric called precision, which is defined as the ratio of relevant items selected to the number of items selected. We focus on the top L predicted links, if there are L_r relevant links (i.e., the links in the probe set), the precision equals L_r/L. Clearly, higher precision means higher prediction accuracy, or that the similarity is quite convincingly reasonable.

2.5 Dimensionality Reduction

High dimensional datasets present many mathematical challenges as well as some opportunities, and are bound to give rise to new theoretical developments. One of the problems with high dimensional datasets is that, in many cases, not all the measured variables are "important" for understanding the underlying phenomena of interest. While certain methods can construct predictive models with high accuracy from high dimensional data, it is still of interest in many applications to reduce the dimension of the original data prior to any modeling of the data.

In machine learning, dimensionality reduction is the process of reducing the number of random variables under consideration, and can be divided into feature selection and feature extraction.

Feature selection approaches try to find a subset of the original variables (also called features or attributes). Two strategies are filter and wrapper approaches. See also combinatorial optimization problems. In some cases, data analysis such as regression or classification can be done in the reduced space more accurately than in the original space.

Feature extraction transforms the data in the high-dimensional space to a space of fewer dimensions. The data transformation may be linear, as in principal component analysis (PCA), but many nonlinear dimensionality reduction techniques also exist.

2.5.1 Principal Component Analysis

Principal component analysis (PCA) is a mathematical procedure that uses an orthogonal transformation to convert a set of observations of possibly correlated variables into a set of values of linearly uncorrelated variables called principal components. The number of principal components is less than or equal to the number of original variables. This transformation is defined in such a way that the first principal component has the largest possible variance (that is, accounts for as much of the variability in the data as possible), and each succeeding component in turn has the highest variance possible under the constraint that it be orthogonal to (i.e., uncorrelated with) the preceding components. Principal components are guaranteed to be independent only if the data set is jointly normally distributed. PCA is sensitive to the relative scaling of the original variables.

Define a data matrix, X^T, with zero empirical mean (the empirical (sample) mean of the distribution has been subtracted from the data set), where each of the n rows represents a different repetition of the experiment, and each of the m columns gives a particular kind of datum (say, the results from a particular probe). (Note that X^T is defined here and not X itself, and what we are calling X^T is often alternatively denoted as X itself.) The singular value decomposition of X is $X = W\Sigma VT$, where the $m \times m$ matrix W is the matrix of eigenvectors of the covariance matrix XX^T, the matrix Σ is an $m \times n$ rectangular diagonal matrix with nonnegative real numbers on the diagonal, and the $n \times n$ matrix V is the matrix of eigenvectors of X^TX. The PCA transformation that preserves dimensionality (that is, gives the same number of principal components as original variables) is then given by:

$$Y^T = X^TW = V\Sigma^TW^TW = V\Sigma^T,$$

V is not uniquely defined in the usual case when $m < n-1$, but Y will usually still be uniquely defined. Since W (by definition of the SVD of a real matrix) is an orthogonal matrix, each row of Y^T is simply a rotation of the corresponding row of X^T. The first column of Y^I is made up of the "scores" of the cases with respect to the "principal" component, the next column has the scores with respect to the "second principal" component, and so on. If we want a reduced-dimensionality representation, we can project X down into the reduced space defined by only the first L singular vectors, W_L:

$$Y = W^T_L X = \Sigma_L V^T,$$

where $\Sigma_L = I_{L \times m}\Sigma$ with $I_{L \times m}$ the $L \times m$ rectangular identity matrix. The matrix W of singular vectors of X is equivalently the matrix W of eigenvectors of the matrix of observed covariances $C = XX^T$,

$$XX^T = W\Sigma\Sigma^TW^T,$$

Given a set of points in Euclidean space, the first principal component corresponds to a line that passes through the multidimensional mean and minimizes the sum of squares of the distances of the points from the line. The second principal component corresponds to the same concept after all correlation with the first principal component has been subtracted from the points. The singular values (in Σ) are the square roots of the eigenvalues of the matrix XX^T. Each eigenvalue is proportional to the portion of the "variance" (more correctly of the sum of the squared distances of the points from their multidimensional mean) that is correlated with each eigenvector. The sum of all the eigenvalues is equal to the sum of the squared distances of the points from their multidimensional mean. PCA essentially rotates the set of points around their mean in order to align with the principal components. This moves as much of the variance as possible (using an orthogonal transformation) into the first few dimensions. The values in the remaining dimensions, therefore, tend to be small and may be dropped with minimal loss of information. PCA is often used in this manner for dimensionality reduction PCA is sensitive to the scaling of the variables. If we have just two variables and they have the same sample variance and are positively correlated, then the PCA will entail a rotation by 45 degrees and the "loadings" for the two variables with respect to the principal component will be equal. But if we multiply all values of the first variable by 100, then the principal component will be almost the same as that variable, with a small contribution from the other variable, whereas the second component will be almost aligned with the second original variable. This means that whenever the different variables have different units (like temperature and mass), PCA is a somewhat arbitrary method of analysis.

2.5.2 Independent Component Analysis

Independent component analysis (ICA) is a computational method for separating a multivariate signal into additive subcomponents supposing the mutual statistical independence of the non-Gaussian source signals. When the independence assumption is correct, blind ICA separation of a mixed signal gives very good results. It is also used for signals that are not supposed to be generated by a mixing for analysis purposes. A simple application of ICA is the "cocktail party problem", where the underlying speech signals are separated from a sample data consisting of people talking simultaneously in a room. Usually the problem is simplified by assuming no time delays or echoes. An important note to consider is that if N sources are present, at least N observations (e.g., microphones) are needed to get the original signals. This constitutes the square case ($J = D$, where D is the input dimension of the data and J is the dimension of the model). Other cases of under-determined ($J < D$) and overdetermined ($J > D$) have been investigated. Linear independent component analysis can be divided into noiseless and noisy cases, where noiseless ICA is a special case of noisy ICA. Nonlinear ICA should be considered as a separate case.

So the general definition is as follows: the data is represented by the random vector $x = (x_1, \cdots, x_m)^T$ and the components as the random vector $s = (s_1, \cdots, s_n)^T$. The task is to transform the observed data x using a linear static transformation W as $s = Wx$ into maximally independent components s measured by some function $F(s_1, \cdots, s_n)$ of independence. In Linear noiseless ICA model, the components of the observed random vector $x = (x_1, \cdots, x_m)^T$ are generated as a sum of the independent components s_k, $k = 1, \cdots, n$:

$$x_i = a_{i,1}s_1 + \cdots + a_{i,k}s_k + \cdots + a_{i,n}s_n$$

weighted by the mixing weights $a_{i,k}$. The same generative model can be written in vectorial form as $x = \sum_{k=1}^{n} s_k a_k$, where the observed random vector x is represented by the basis vectors $a_k = (a_{1,k}, \cdots, a_{m,k})^T$. The basis vectors a_k form the columns of the mixing matrix $A = (a_1, \cdots, a_n)$ and the generative formula can be written as $x = As$, where $s = (s_1, \cdots, s_n)^T$. Given the model and realizations (samples) $x_1, \cdots, x_N$ of the random vector x, the task is to estimate both the mixing matrix A and the sources s. This is done by adaptively calculating the vectors ω and setting up a cost function which either maximizes the non-gaussianity of the calculated $s_k = (\omega^T \times x)$ or minimizes the mutual information. In some cases, a prior knowledge of the probability distributions of the sources can be used in the cost function. The original sources s can be recovered by multiplying the observed signals x with the inverse of the mixing matrix $W = A^{-1}$, also known as the unmixing matrix. Here it is assumed that the mixing matrix is square ($n = m$). If the number of basis vectors is greater than the dimensionality of the observed vectors, $n > m$, the task is overcomplete but is still solvable with the pseudo inverse. In Linear noisy ICA model, with the added assumption of zeromean and uncorrelated Gaussian noise $n \sim N(0, \text{diag}(\Sigma))$, the ICA model takes the form $x = As + n$. And in Non-linear ICA model, the mixing of the sources does not need to be linear. Using a nonlinear mixing function $f(\cdot \mid \theta)$ with parameters θ, non-linear ICA model is $x = f(s \mid \theta) + n$.

2.5.3 Non-negative Matrix Factorization

Non-negative matrix factorization (NMF) is a group of algorithms in multivariate analysis and linear algebra where a matrix, X, is factorized into (usually) two matrices, W and H: $\text{nmf}(X) \rightarrow WH$.

Factorization of matrices is generally non-unique, and a number of different methods of doing so have been developed (e.g., principal component analysis and singular value decomposition) by incorporating different constraints; non-negative matrix factorization differs from these methods in that it enforces the constraint that the factors W and H must be non-negative, i.e., all elements must be equal to or greater than zero.

Let matrix V be the product of the matrices W and H such that:

$$WH = V$$

Matrix multiplication can be implemented as linear combinations of column vectors in W with coefficients supplied by cell values in H. Each column in V can be computed as follows:

$$v_i = \sum_{j=1}^{N} H_{ji} w_j$$

where N is the number of columns in W, v_i is the ith column vector of the product matrix V, H_{ji} is the cell value in the jth row and ith column of the matrix H, w_j is the jth column of the matrix W. When multiplying matrices the factor matrices can be of significantly lower rank than the product matrix and it is this property that forms the basis of NMF. If we can factorize a matrix into factors of significantly lower rank than the original matrix, then the column vectors of the first factor matrix can be considered as spanning vectors of the vector space defined by the original matrix.

Here is an example based on a text-mining application:

- Let the input matrix (the matrix to be factored) be V with 10000 rows and 500 columns where words are in rows and documents are in columns. In other words, we have 500 documents indexed by 10000 words. It follows that a column vector v in V represents a document.
- Assume we ask the algorithm to find 10 features in order to generate a features matrix W with 10000 rows and 10 columns and a coefficients matrix H with 10 rows and 500 columns.
- The product of W and H is a matrix with 10000 rows and 500 columns, the same shape as the input matrix V and, if the factorization worked, also a reasonable approximation to the input matrix V.
- From the treatment of matrix multiplication above it follows that each column in the product matrix WH is a linear combination of the 10 column vectors in the features matrix W with coefficients supplied by the coefficients matrix H.

This last point is the basis of NMF because we can consider each original document in our example as being built from a small set of hidden features. NMF generates these features.

It is useful to think of each feature (column vector) in the features matrix W as a document archetype comprising a set of words where each word's cell value defines the word's rank in the feature: The higher a word's cell value the higher the word's rank in the feature. A column in the coefficients matrix H represents an original document with a cell value defining the document's rank for a feature. This follows because each row in H represents a feature. We can now reconstruct a document (column vector) from our input matrix by a linear combination of our features (column vectors in W) where each feature is weighted by the feature's cell value from the document's column in H.

2.5.4 Singular Value Decomposition

A number of data sets are naturally described in matrix form. Examples range from microarrays to collaborative filtering data, to the set of pairwise distances of a cloud of points. In many of these examples, singular value decomposition (SVD) provides an efficient way to construct a low-rank approximation thus achieving both dimensionality reduction, and effective denoizing. SVD is also an important tool in the design of approximate linear algebra algorithms for massive data sets.

In linear algebra, the singular value decomposition (SVD) is a factorization of a real or complex matrix, with many useful applications in signal processing and statistics. Formally, the singular value decomposition of an mn real or complex matrix M is a factorization of the form

$$M = U\Sigma V^*,$$

where U is an $m \times m$ real or complex unitary matrix, is an $m \times n$ rectangular diagonal matrix with non-negative real numbers on the diagonal, and V^* (the conjugate transpose of V) is an $n \times n$ real or complex unitary matrix. The diagonal entries $\Sigma_{i,i}$ of Σ are known as the singular values of M. The m columns of U and the n columns of V are called the left-singular vectors and right-singular vectors of M, respectively.

The singular value decomposition and the eigendecomposition are closely related. Namely:

- The left-singular vectors of M are eigenvectors of MM^*
- The right-singular vectors of M are eigenvectors of MM^*
- The non-zero-singular values of M (found on the diagonal entries of Σ) are the square roots of the non-zero eigenvalues of both M^*M and MM^*.

Applications which employ the SVD include computing the pseudo inverse, least squares fitting of data, matrix approximation, and determining the rank, range and null space of a matrix.

2.6 Chapter Summary

In this chapter, we have systematically presented the mathematical foundations used in this book. We start with data organization and distribution followed by the intensive discussion on distance and similarity measures. This chapter also covers the important issue of dimensionality reduction approaches that is commonly used in vector space models. The aim of this chapter is to lay down a solid foundation for readers to better understand the techniques and algorithms mentioned in later chapters.

References

[1] Chandrika Kamath. *Scientific Data Mining: A Practical Perspective*, Lawrence Livermore National Laboratory, Livermore, California, 2009.

[2] S. Brin and L. Page. *Anatomy of a Large-scale Hypertextual Web Search engine*, Proc. 7th Intl. World-Wide-Web Conference, pp. 107C117, 1998.

[3] Norman Lloyd Johnson, Samuel Kotz, N. Balakrishnan. *Continuous Univariate Distributions*, John Wiley and Sons, 2005.

[4] N. Samuel Kotz, Balakrishnan and Norman L. Johnson. *Continuous Multivariate Distributions, Models and Applications*, John Wiley and Sons, 2000.

[5] Anand Rajaraman and Jeffrey D. Ullman. *Mining of Massive Datasets*, Palo Alto, CA, pp. 74C79, 2011.

[6] J. C. Christopher, Burges. *Dimension Reduction*, Now Publishers Inc., 2010.

CHAPTER 3
Data Preparation

Data preparation is the start of the data mining process. The data mining results heavily rely on the data quality prepared before the mining process. It is a process that involves many different tasks and which cannot be fully automated. Many of the data preparation activities are routine, tedious, and time consuming. It has been estimated that data preparation accounts for 60 percent to 80 percent of the time spent on a data mining project. Figure 3.0.1 shows the main steps of data mining. From the figure, we can see that the data preparation takes an important role in data mining.

Data preparation is essential for successful data mining. Poor quality data typically result in incorrect and unreliable data mining results. Data preparation improves the quality of data and consequently helps improve the quality of data mining results. The well known saying "garbage-in garbage-out" is very relevant to this domain. This chapter contributes to the related topics with respect to data preparation, covering attribute selection, data cleaning and integrity, multiple model integration and so on.

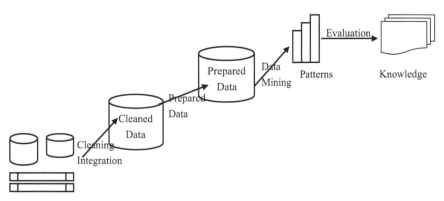

Figure 3.0.1: Main steps of data mining

3.1 Attribute Selection

3.1.1 Feature Selection

Feature selection, also known as variable selection, attribute reduction, feature selection or variable subset selection, is the technique of selecting a subset of relevant features for building robust learning models. Attribute selection is a particularly important step in analyzing the data from many experimental techniques in biology, such as DNA microarrays, because they often entail a large number of measured variables (features) but a very low number of samples. By removing most irrelevant and redundant features from the data, feature selection helps improve the performance of learning models by:

- Alleviating the effect of the curse of dimensionality
- Enhancing generalization capability
- Speeding up learning process
- Improving model interpretability.

Feature selection also helps people to acquire better understanding about their data by telling them which are the important features and how they are related with each other.

Simple feature selection algorithms are ad hoc, but there are also more methodical approaches. From a theoretical perspective, it can be shown that optimal feature selection for supervised learning problems requires an exhaustive search of all possible subsets of features of the chosen cardinality. If large numbers of features are available, this is impractical. For practical supervised learning algorithms, the search is for a satisfactory set of features instead of an optimal set. Feature selection algorithms typically fall into two categories: feature ranking and subset selection. Feature ranking ranks the features by a metric and eliminates all features that do not achieve an adequate score. Subset selection searches the set of possible features for the optimal subset. In statistics, the most popular form of feature selection is stepwise regression. It is a greedy algorithm that adds the best feature (or deletes the worst feature) at each round. The main control issue is deciding when to stop the algorithm. In machine learning, this is typically done by cross-validation. In statistics, some criteria are optimized. This leads to the inherent problem of nesting. More robust methods have been explored, such as branch and bound and piecewise linear network.

3.1.1.1 Subset Selection

Subset selection evaluates a subset of features as a group for suitability. Subset selection algorithms can be broken into Wrappers, Filters and

Embedded. Wrappers use a search algorithm to search through the space of possible features and evaluate each subset by running a model on the subset. Wrappers can be computationally expensive and have a risk of over fitting to the model. Filters are similar to Wrappers in the search approach, but instead of evaluating against a model, a simpler filter is evaluated. Embedded techniques are embedded in and specific to a model. Many popular search approaches use greedy hill climbing, which iteratively evaluates a candidate subset of features, then modifies the subset and evaluates if the new subset is an improvement over the old. Evaluation of the subsets requires a scoring metric that grades a subset of features. Exhaustive search is generally impractical, so at some implementor (or operator) defined stopping point, the subset of features with the highest score discovered up to that point is selected as the satisfactory feature subset. The stopping criterion varies by algorithm; possible criteria include: a subset score exceeds a threshold, a program's maximum allowed run time has been surpassed, etc. Alternative search-based techniques are based on targeted projection pursuit which finds low-dimensional projections of the data that score highly: the features that have the largest projections in the lower dimensional space are then selected. Search approaches include:

- Exhaustive
- Best first
- Simulated annealing
- Genetic algorithm
- Greedy forward selection
- Greedy backward elimination
- Targeted projection pursuit
- Scatter search
- Variable neighborhood search.

Two popular filter metrics for classification problems are correlation and mutual information, although neither are true metrics or 'distance measures' in the mathematical sense, since they fail to obey the triangle inequality and thus do not compute any actual 'distance'—they should rather be regarded as 'scores'. These scores are computed between a candidate feature (or set of features) and the desired output category. There are, however, true metrics that are a simple function of the mutual information; see here. Other available filter metrics include:

- Class reparability
- Error probability
- Inter-class distance
- Probabilistic distance
- Entropy

- Consistency-based feature selection
- Correlation-based feature selection.

3.1.1.2 Optimality Criteria

There are a variety of optimality criteria that can be used for controlling feature selection. The oldest are Mallows' Cp statistic and Akaike information criterion (AIC). These add variables if the t-statistic is bigger than $\sqrt{2}$. Other criteria are Bayesian information criterion (BIC) which uses $\sqrt{\log n}$, minimum description length (MDL) which asymptotically uses $\sqrt{\log n}$, Bonnferroni/RIC which use $\sqrt{2 \log p}$, maximum dependency feature selection, and a variety of new criteria that are motivated by false discovery rate (FDR) which use something close to $\sqrt{2 \log \frac{p}{q}}$.

3.1.1.3 Correlation Feature Selection

The Correlation Feature Selection (CFS) measure evaluates subsets of features on the basis of the following hypothesis: "Good feature subsets contain features highly correlated with the classification, yet uncorrelated to each other [2]. The following equation gives the merit of a feature subset S consisting of k features: $Merit_{S_K} = \dfrac{k\bar{r}_{cf}}{\sqrt{k+k(k-1)\bar{r}_{ff}}}$, where $\bar{r}_{cf}$ is the average value of all feature-classification correlations, and $\bar{r}_{ff}$ is the average value of all feature-feature correlations.

The CFS criterion is defined as follows:

$$CFS = \max_{S_k} \left[\frac{r_{cf1} + r_{cf2} +, ..., + r_{cfk}}{\sqrt{k + 2(r_{f1f1} + ... + r_{fifj} + ... + r_{fkfk})}} \right].$$

The r_{cf_i} and r_{fifj} variables are referred to as correlations, but are not necessarily Pearson's correlation coefficient or Spearman's. Dr Mark Hall's dissertation uses neither of these, but uses three different measures of relatedness, minimum description length (MDL), symmetrical uncertainty, and relief.

Let x_i be the set membership indicator function for feature f_i; then the above can be rewritten as an optimization problem:

$$CFS = \max_{x \in \{0.1\}^n} \left[\frac{(\sum_{i=1}^{n} a_i x_i)}{\sum_{i=1}^{n} x_i + \sum_{i \neq j} 2b_{ij} x_i x_j} \right].$$ The combinatorial problems above are, in fact, mixed 0–1 linear programming problems that can be solved by using branch-and-bound algorithms.

3.1.1.4 Software for Feature Selection

Many standard data analysis software systems are often used for feature selection, such as SciLab, NumPy and the *R* language. Other software systems are tailored specifically to the feature-selection task [1]:

- Weka—freely available and open-source software in Java.
- Feature Selection Toolbox 3—freely available and open-source software in C++.
- RapidMiner—freely available and open-source software.
- Orange—freely available and open-source software (module orngFSS).
- TOOLDIAG Pattern recognition toolbox - freely available C toolbox.
- Minimum redundancy feature selection tool - freely available C/Matlab codes for selecting minimum redundant features.
- A C# Implementation of greedy forward feature subset selection for various classifiers (e.g., LibLinear, SVM-light).
- MCFS-ID (Monte Carlo Feature Selection and Interdependency Discovery) is a Monte Carlo method-based tool for feature selection. It also allows for the discovery of interdependencies between the relevant features. MCFS-ID is particularly suitable for the analysis of high-dimensional, ill-defined transactional and biological data.
- RRF is an R package for feature selection and can be installed from R. RRF stands for Regularized Random Forest, which is a type of Regularized Trees. By building a regularized random forest, a compact set of non-redundant features can be selected without loss of predictive information. Regularized trees can capture non-linear interactions between variables, and naturally handle different scales, and numerical and categorical variables.

3.1.2 Discretizing Numeric Attributes

We can turn a numeric attribute into a nominal/categorical one by using some sort of discretization. This involves dividing the range of possible values into sub-ranges called buckets or bins. For example: an age attribute could be divided into these bins:

child: 0–12
teen: 12–17
young: 18–35
middle: 36–59
senior: 60–
What if we don't know which sub-ranges make sense? [5, 7, 6]

- Equal-width binning divides the range of possible values into N subranges of the same size and bin width = (max value - min value)$/N$

For example: if the observed values are all between 0 and 100, we could create 5 bins as follows:

(1) width = (100–0)/5 = 20
(2) bins: [0–20], (20–40], (40–60], (60–80], (80–100] [or] means the endpoint is included (or) means the endpoint is not included
(3) typically, the first and last bins are extended to allow for values outside the range of observed values (-infinity-20], (20–40], (40–60], (60–80], (80-infinity).

- Equal-frequency or equal-height binning divides the range of possible values into N bins, each of which holds the same number of training instances.

For example: let's say we have 10 training examples with the following values for the attribute that we are discrediting: 5, 7, 12, 35, 65, 82, 84, 88, 90, 95 to create 5 bins, we would divide up the range of values so that each bin holds 2 of the training examples: 5–7, 12–35, 65–82, 84–88, 90–95. To select the boundary values for the bins, this method typically chooses a value halfway between the training examples on either side of the boundary. For example: (7 + 12)/2 = 9.5 (35 + 65)/2 = 50

3.2 Data Cleaning and Integrity

3.2.1 Missing Values

Imagine that you need to analyze All Electronics sales and customer data. You note that many tuples have no recorded value for several attributes, such as customer income. How can you go about filling in the missing values for this attribute? Let's look at the following methods [9, 8]:

- Ignore the tuple: This is usually done when the class label is missing (assuming the mining task involves classification or description). This method is not very effective, unless the tuple contains several attributes with missing values. It is especially poor when the percentage of missing values per attribute varies considerably.
- Fill in the missing value manually: In general, this approach is time-consuming and may not be feasible given a large data set with many missing values.
- Use a global constant to fill in the missing value: Replace all missing attribute values by the same constant, such as a label like "Unknown". If missing values are replaced by, say, "Unknown", then the mining

program may mistakenly think that they form an interesting concept, since they all have a value in common—that of "Unknown". Hence, although this method is simple, it is not recommended.

- Use the attribute mean to fill in the missing value: For example, suppose that the average income of All Electronics customers is $28,000. Use this value to replace the missing value for income.
- Use the attribute mean for all samples belonging to the same class as the given tuple: For example, if classifying customers according to credit_risk, replace the missing value with the average income value for customers in the same credit_risk category as that of the given tuple.
- Use the most probable value to fill in the missing value: This may be determined with inference-based tools using a Bayesian formalism or decision tree induction. For example, sing the other customer attributes in your data set, you may construct a decision tree to predict the missing values for income.

Methods 3 to 6 bias the data. The filled-in value may not be correct. Method 6, however, is a popular strategy. In comparison to the other methods, it uses the most information from the present data to predict missing values.

3.2.2 Detecting Anomalies

Anomaly detection, also referred to as outlier detection [2], refers to detecting patterns in a given data set that do not conform to an established normal behavior. The patterns thus detected are called anomalies and often translate to critical and actionable information in several application domains. Anomalies are also referred to as outliers, change, deviation, surprise, aberrant, peculiarity, intrusion, etc. In particular in the context of abuse and network intrusion detection, the interesting objects are often not rare objects, but unexpected bursts of activity. This pattern does not adhere to the common statistical definition of an outlier as a rare object, and many outlier detection methods (in particular unsupervised methods) will fail on such data, unless it has been aggregated appropriately. Instead, a cluster analysis algorithm may be able to detect the micro clusters formed by these patterns. Three broad categories of anomaly detection techniques exist. *Unsupervised anomaly detection techniques* detect anomalies in an unlabeled test data set under the assumption that the majority of the instances in the data set are normal by looking for instances that seem to fit least to the remainder of the data set. *Supervised anomaly detection techniques* require a data set that has been labeled as "normal" and "abnormal" and involves training a classifier (the key difference to many other statistical classification problems is the inherent unbalanced nature of outlier detection). *Semi-supervised anomaly detection techniques* construct a model representing normal behavior from a given normal training data set, and then testing the likelihood of a test instance to be generated by the learnt model.

3.2.3 Applications

Anomaly detection is applicable in a variety of domains, such as intrusion detection, fraud detection, fault detection, system health monitoring, event detection in sensor networks, and detecting eco-system disturbances. It is often used in preprocessing to remove anomalous data from the dataset. In supervised learning, removing the anomalous data from the dataset often results in a statistically significant increase in accuracy.

(1) Popular techniques

Several anomaly detection techniques have been proposed in literature. Some of the popular techniques are:

- Distance based techniques (k-nearest neighbor, Local Outlier Factor)
- One class support vector machines
- Replicator neural networks
- Cluster analysis based outlier detection
- Pointing at records that deviate from association rules
- Conditional anomaly concept.

(2) Application to data security

Anomaly detection was proposed for Intrusion Detection Systems (IDS) by Dorothy Denning in 1986. Anomaly detection for IDS is normally accomplished with thresholds and statistics, but can also be done with soft computing and inductive learning. Types of statistics proposed by 1999 included profiles of users, workstations, networks, remote hosts, groups of users, and programs based on frequencies, means, variances, covariances, and standard deviations. The counterpart of Anomaly Detection in Intrusion Detection is Misuse Detection.

(3) Time series outlier detection

Parametric tests to find outliers in time series are implemented in almost all statistical packages: Demetra+, for example, uses the most popular ones. One way to detect anomalies in time series is a simple non-parametric method called *washer*. It uses a non-parametric test to find one or more outliers in a group of even very short time series. The group must have a similar behaviour, as explained more fully below. An example is that of municipalities cited in the work of Dahlberg and Johanssen (2000). Swedish municipalities expenditures between 1979 and 1987 represent 256 time series. If you consider three years such as, for example, 1981,1982 and 1983, you have 256 simple polygonal chains made of two lines segments. Every couple of segments can approximate a straight line or a convex downward (or convex upward) simple polygonal chain. The idea is to find outliers among the couples of segments that performs in a too much different way from the other couples. In the washer procedure every couple of segments is represented by an index and a non-parametric test (Sprent test) is applied to the unknown distribution of those indices. For implementing washer methodology you can download an open source R (programming language) function with a simple numeric example.

3.3 Multiple Model Integration

3.3.1 Data Federation

Data federation is a brand new idea for integration of data from many diffract sources. Many organizations and companies store their data in different ways, like transactional databases, data warehouses, business intelligence systems, legacy systems and so on. The problem arises, when someone needs to access data from some of these sources [8, 4, 3]. There is no easy way to retrieve the data, because every storage system has its own way of accessing it. In order to help getting to the data from many sources, there are some ways to integrate the data, and the most advanced of them is *data federation*. To integrate the data it has to be copied and moved, because the integrated data need to be kept together. Of course it has its defects, like the time needed to copy and move the data, and some copyright infringements during copying. The data also occupied more disk space than it actually needed, because it was kept in few instances. There were also some problems with data refreshing, because if there was more than one instance of the data, only the modified instance was up to date, so all others instances of the data has to be refreshed. Of course it slowed down the integration system. In response to these problems, the IT specialists created a new data integration system called *data federation*. The idea of data federation is to integrate data from many individual sources and make access to them as easy as possible. The target has to be reached without moving or copying the data. In fact, the data sources can be in any location. It only has to be online. Also, every data source can be made using different technology, standard and architecture. For the end user it will feel like one big data storage system. The data federation supports many data storage standards. From the SQL relational databases like Mysql, PostgreSQL, InterBase, IBM DB2, Firebird and Oracle through directory services and object-based databases like LDAP and OpenLDAP, to data warehouses and Business-Intelligence systems. The goal is to make the data federation system work with every standard that is used to store data in companies and other organizations. The data that is already integrated with the data federation system is called a federated database or a virtual database. Federated database allows users to read and write the data without even knowing that it comes from many different sources. The user doesn't need to know how to use a database system, or how to access data in a directory service. All he needs is to know is how to use the unified front-end of the data federation system. The data federation system in many cases might be the best way to unify the data kept in different places in many different ways. It's simple, easy for the end users, and an efficient solution that will make accessing the data a lot easier.

3.3.2 Bagging and Boosting

(1) Bagging

The concept of bagging (voting for classification, averaging for regression-type problems with continuous dependent variables of interest) applies to the area of predictive data mining, to combine the predicted classifications (prediction) from multiple models, or from the same type of model for different learning data. It is also used to address the inherent instability of results when applying complex models to relatively small data sets. Suppose your data mining task is to build a model for predictive classification, and the dataset from which to train the model (learning data set, which contains observed classifications) is relatively small. You could repeatedly sub-sample (with replacement) from the dataset, and apply, for example, a tree classifier (e.g., C&RT and CHAID) to the successive samples. In practice, very different trees will often be grown for the different samples, illustrating the instability of models often evident with small datasets. One method of deriving a single prediction (for new observations) is to use all trees found in the different samples, and to apply some simple voting: The final classification is the one most often predicted by the different trees. Note that some weighted combination of predictions (weighted vote, weighted average) is also possible, and commonly used. A sophisticated (machine learning) algorithm for generating weights for weighted prediction or voting is the Boosting procedure.

(2) Boosting

The concept of boosting applies to the area of predictive data mining, to generate multiple models or classifiers (for prediction or classification), and to derive weights to combine the predictions from those models into a single prediction or predicted classification (see also Bagging). A simple algorithm for boosting works like this: Start by applying some method (e.g., a tree classifier such as C&RT or CHAID) to the learning data, where each observation is assigned an equal weight. Compute the predicted classifications, and apply weights to the observations in the learning sample that are inversely proportional to the accuracy of the classification. In other words, assign greater weight to those observations that were difficult to classify (where the misclassification rate was high), and lower weights to those that were easy to classify (where the misclassification rate was low). In the context of C&RT for example, different misclassification costs (for the different classes) can be applied, inversely proportional to the accuracy of prediction in each class. Then apply the classifier again to the weighted data (or with different misclassification costs), and continue with the next iteration (application of the analysis method for classification to the re-weighted data). Boosting will generate a sequence of classifiers, where each consecutive classifier in the sequence is an "expert" in classifying observations that were not well classified by those preceding it. During deployment (for prediction or classification of new cases), the predictions from the different classifiers can then be combined (e.g., via voting, or some weighted voting procedure) to derive a single best prediction or classification. Note that boosting can also be applied to learning methods that do not explicitly support weights or misclassification costs. In that case, random sub-sampling can be applied to the learning data in the successive steps of the iterative boosting procedure, where the probability for selection of an observation into the subsample is inversely proportional to the accuracy of the prediction for that observation in the previous iteration (in the sequence of iterations of the boosting procedure).

3.4 Chapter Summary

In this section, we summarize the techniques involved in data preparation, which is an essential step for the success of data mining. Particularly, we discuss the issues of feature selection, data cleaning, missing values and data federation.

References

[1] http://en.wikipedia.org/wiki/Feature_selection# Correlation feature selection.
[2] M.A. Hall and L.A. Smith. Feature selection for machine learning: Comparing a correlation-based filter approach to the wrapper. *In: FLAIRS Conference*, pp. 235–39, 1999.
[3] Z. Huang, M.-L. Shyu and J. M. Tien. Multi-model integration for long-term time series prediction. *In: IRI'12*, pp. 116–23, 2012.
[4] A. Lazarevic, A. Lazarevic and Z. Obradovic. Data reduction using multiple models integration, 2000.

[5] H. Liu and R. Setiono. Chi2: Feature selection and discretization of numeric attributes. *In: Proceedings of the Seventh International Conference on Tools with Artificial Intelligence,* pp. 388–391, 1995.

[6] H. Liu and R. Setiono. Chi2: Feature selection and discretization of numeric attributes. *Tools with Artificial Intelligence, IEEE International Conference on,* 0: 388, 1995.

[7] H. Liu and R. Setiono. Feature selection via discretization. *IEEE Transactions on Knowledge and Data Engineering,* 9: 642–645, 1997.

[8] J. I. Maletic and A. Marcus. *Data Cleansing: Beyond Integrity Analysis,* 2000.

[9] E. Rahm and H. H. Do. *Data Cleaning: Problems and Current Approaches. IEEE Data Engineering Bulletin,* 23: 2000, 2000.

CHAPTER 4

Clustering Analysis

4.1 Clustering Analysis

Clustering analysis is an important learning method which doesn't need any prior knowledge. Clustering is usually performed when no information is available concerning the membership of data items to predefined classes. For this reason, clustering is traditionally seen as part of unsupervised learning. We nevertheless speak here of unsupervised clustering to distinguish it from a more recent and less common approach that makes use of a small amount of supervision to "guide" or "adjust" clustering. In this chapter, we focus on discussing this unsupervised learning method. The aim of clustering analysis is to divide data into groups (clusters) that are meaningful, useful or both. For meaningful groups, the goal of clustering is to capture the natural structure of data. In some cases, however, clustering is only a useful starting point for other purposes, such as data summarization. Whether for understanding or utility, clustering analysis has played an important role in a wide variety of fields: computer science, pattern recognition, information retrieval, machine learning, biology, data mining etc. Many data mining queries are concerned either with how the data objects are grouped or which objects could be considered remote from natural groupings. There have been many works on cluster analysis, but we are now witnessing a significant resurgence of interest in new clustering techniques. Scalability and high dimensionality are not the only focus of the recent research in clustering analysis. Indeed, it is getting difficult to keep track of all the new clustering strategies, their advantages and shortcomings. The following are the typical requirements for a good clustering technique in data mining [30, 29]:

- *Scalability*: The cluster method should be applicable to huge databases and performance should decrease linearly with data size increase.
- *Versatility*: Clustering objects could be of different types—numerical data, boolean data or categorical data. Ideally a clustering method should be suitable for all different types of data objects.

- *Ability to discover clusters with different shapes*: This is an important requirement for spatial data clustering. Many clustering algorithms can only discover clusters with spherical shapes.
- *Minimal input parameter*: This method should require a minimum amount of domain knowledge for correct clustering. However, most current clustering algorithms have several key parameters and are thus not practical for use in real world applications.
- *Robust with regard to noise*: This is important because noise exists everywhere in practical problems. A good clustering algorithm should be able to perform successfully even in the presence of a great deal of noise.
- *Insensitive to the data input order*: The clustering method should give consistent results irrespective of the order the data is presented.
- *Scaleable to high dimensionality*: The ability to handle high dimensionality is very challenging but real data sets are often multidimensional.

There is no single algorithm that can fully satisfy all the above requirements. It is important to understand the characteristics of each algorithm so that the proper algorithm can be selected for the clustering problem at hand. Recently, there are several new clustering techniques offering useful advances, possibly even complete solutions. During the past decades, clustering analysis has been used to deal with practical problems in many applications, as summed up by Han [30, 29]. Biology. Biologists have spent many years creating a taxonomy (hierarchical classification) of all living things: kingdom, class, order, family, genus and species. More recently, biologists have applied clustering to analyze the large amounts of genetic information that are now available. For example, clustering has been used to find groups of genes that have similar functions from high dimensional genes data.

It has been used for *Information Retrieval*. The World Wide Web consists of billions of Web pages, and the results of a query to a search engine can return thousands of pages. Clustering can be used to group these search results into a small number of clusters, each of which captures a particular aspect of the query.

Climate. Understanding the earth's climate requires finding patterns in the atmosphere and ocean. To that end, clustering analysis has been applied to find patterns in the atmospheric pressure of polar regions and areas of the ocean that have a significant impact on land climate.

Psychology and Medicine. All illness or condition frequently has a number of variations, and cluster analysis can be used to identify these different subcategories. For example, clustering has been used to identify types of depression. Cluster analysis can also be used to detect patterns in the spatial or temporal distribution of a disease.

Business. Businesses collect large amounts of information on current and potential customers. Clustering can be used to segment customers into a small number of groups for additional analysis and marketing activities.

This chapter provides an introduction to clustering analysis. We begin with the discussion of data types which have been met in clustering analysis, and then, we will introduce some traditional clustering algorithms which have the ability to deal with low dimension data clustering. High-dimensional problem is a new challenge for clustering analysis, and lots of high-dimensional clustering algorithms have been proposed by researchers. Constraint-based clustering algorithm is a kind of semi-supervised learning method, and it will be briefly discussed in this chapter as well. Consensus cluster algorithm focuses on the clustering results derived by other traditional clustering algorithms. It is a new method to improve the quality of clustering result.

4.2 Types of Data in Clustering Analysis

As we know, clustering analysis methods could be used in different application areas. So for clustering, different types of data sets will be met. Data sets are made up of data objects (also referred to as samples, examples, instance, data points, or objects) and a data object represents an entity. For example, in a sales database, the objects may be customers, store items and sales; in a medical database, the objects may be patients; in a university database, the objects may be students, course, professor, salary; in a webpage database, the objects maybe the users, links and pages; in a tagging database, the objects may be users, tags and resources, and so on. In clustering scenario, there have two traditional ways to organize the data objects: Data Matrix and Proximity Matrix.

4.2.1 Data Matrix

A set of objects is represented as an m by n matrix, where there are m rows, one for each object, and n columns, one for each attribute. This matrix has different names, e.g., pattern matrix or data matrix, depending on the particular field. Figure 4.2.1 below, provides a concrete example of web usage data objects and their corresponding data matrix, where s_i, $i=1,...,m$ indicates m user sessions and p_j, $j=1,...,n$ indicates n pages, $a_{ij}=1$ indicates s_i has visited pj, otherwise, $a_{ij}=0$. Because different attributes may be measured on different scales, e.g., centimeter and kilogram, the data is sometimes transformed before being used. In cases where the range of

	p_1	p_2	...	p_n
S_1				
S_2		a_{ij}		
...				
S_m				

Figure 4.2.1: An example of data matrix

values differs widely from attribute to attribute, these differing attribute scales can dominate the results of the cluster analysis and it is common to standardize the data so that all attributes are on the same scale. In order to introduce these approaches clearly, we denote that x_i is the i-th data object, x_{ij} is the value of the j-th attribute of the i-th object, and x'_{ij} is the standardized attribute value. There have some common approaches for data standardization as follows:

(1) $x'_{ij} = \dfrac{x_{ij}}{\max\limits_{i} | x_{ij} |}$. Divide each attribute value of an object by the

maximum observed absolute value of that attribute. This restricts all attribute values to lie between -1 and 1. If all the values are positive, all transformed values lie between 0 and 1. This approach may not produce good results unless the attributes are uniformly distributed, and this approach is also sensitive to outliers.

(2) $\dfrac{x'_{ij} = x_{ij} - \mu_j}{\sigma_j}$. For each attribute value subtract off the mean of that

attribute and then divide it by the standard deviation of the attribute, where $\mu_j = \dfrac{1}{m}\sum_{i=1}^{m} x_{ij}$ is the mean of the j-th feature, and $\sigma_j = \dfrac{1}{m}$

$\sqrt{\sum_{i=1}^{m}(x_{ij} - \mu_j)^2}$ is the standard deviation of the j-th feature.

Kaufman et al. indicate that if the data are normally distributed, then most transformed attribute values will be lie between -1 and 1. This approach has no request for the data distribution, but it is also sensitive to the outliers like the first approach.

(3) $\dfrac{x'_{ij} = x_{ij} - \mu'_j}{\sigma^A_j}$. For each attribute value subtract off the mean of that

attribute and divide i+ by the attribute's absolute deviation (KR 90), where $\sigma^A_j = \dfrac{1}{m}\sqrt{\sum_{i=1}^{m} | x_{ij} - \mu_j |}$ is the absolute standard deviation of the

j-th attribute. Typically, most attribute values will lie between -1 and 1. This approach is the most robust in the presence of outliers among three approaches.

4.2.2 The Proximity Matrix

While cluster analysis sometimes uses the original data matrix, many clustering algorithms use a similarity matrix, S, or a dissimilarity matrix, D. For convenience, both matrices are commonly referred to as a proximity matrix, P. A proximity matrix, P, is an m by m matrix containing all the pairwise dissimilarities or similarities between the objects being considered. If x_i and x_j are the i-th and j-th objects respectively, then the entry of p_{ij} is the similarity, s_{ij}, or the dissimilarity, d_{ij}, between x_i and x_j, where, p_{ij} denotes the element at the i-th row and j-th column of the proximity matrix P. For simplicity, we will use p_{ij} to represent either s_{ij} or d_{ij}. Figure 4.2.2 gives an example of proximity matrix in social tagging mining, where the element in the matrix represents the cosine similarity between tags. From the description of data objects, we could see the fact that data objects

1				
0.41	1			
0	0.5	1		
0.58	0	0	1	
0.41	0.50	0.50	0.71	1

Figure 4.2.2: An example of Proximity Matrix

are typically described by attribute. An attributes is a data field which represents a characteristic or a feature of a data object. In the proposed literature, the nouns attribute, dimension, feature, and variable are often used interchangeably. The term "dimension" is commonly used in data warehousing, while the term "feature" is tended to be used in machine learning. For the term "variable", it used to be occurred in statisticians. In data mining and database area, the researchers prefer the term "attribute". For example, attributes describing a user object in a webpage database can include, *user_iD, page_1, page_2, ···, page_n*. A set of attributes used to describe a given object is called an attribute *vector* or *feature vector*. The distribution of data involving one attribute is called *univariate*, and a *bivariate* distribution involves two attributes, and so on. The type of an attribute is determined by the possible values, that is, nominal, binary, ordinal or numeric.

Nominal Attributes. The values of enumerations attribute are symbols or names of things, that is, enumerations is related to names. Each value represents some kind of category, code, or state. There is no meaningful order for the value of these kind of attributes. As we know, in some cases, the values are also known as enumeration attributes. For example, suppose that course is an attribute describing student objects in a university database. The possible values for course are software engineering, mathematics, English writing, and so on. Another example of an enumeration attributes is the color attribute, the possible values of it are red, black, blue, and so on. As we said earlier, the values of enumeration attributes are symbols, however, it is possible to represent such symbols with numbers. In order to achieve this goal, we can assign a code of 0 to red, 1 to black, 2 to blue, and so on.

Binary Attributes. During the application, some binary attributes will be accounted. A binary attribute is a nominal attribute with only two states: 0 or 1, where 0 indicates the attribute is absent and 1 means that it is present. In some sense, binary attributes are referred to as Boolean if the two states correspond to false and true. There are two types of binary attributes: symmetric and asymmetric. For symmetric binary attributes, the states are equally valuable and can carry the same weight, that is, there is no preference on which outcome should be coded as 0 or 1. For instance, the attribute gender having the states male and female. For asymmetric binary attributes, the outcomes of the states are not equally important, such as the positive and negative outcomes of a medical test for HIV.

Ordinal Attributes. An ordinal attribute is an attribute with possible values that have a meaningful order among them, but the magnitude between successive values is not known [30, 29]. Take for example, the grade attribute for a student's test score. The possible values of grade could be A+, A, A-, B+ and so on. Ordinal attributes are useful for registering subjective assessments of qualities that cannot be measured objectively. Thus, ordinal attributes are often used in surveys for rating. For example, in social network area, the participants were asked to rate how good was a movie which they have seen. The rating of the movie had the following categories: 0: excellent, 1: good, 2: normal, 3: bad, 4: very bad. In some cases, ordinal attributes may also be obtained from the discrimination of numeric quantities by splitting the value range into a finite number of ordered categories.

Numeric Attributes. A numeric attribute is represented in integer or real value and it is quantitative. There are two types of numeric attributes: interval-scaled and ratio-scaled. Interval-scaled attributes are measured on a scale of equal size units. The values of interval-scaled attributes have order and can be positive, 0, or negative. In other words, such attributes allow us to compare and quantify the difference between values. For instance, height is a numeric attribute of a person, the height value of Tom is 175 cm and that of Jack is 185 cm. We can then say that Jack is taller than Tom. Ratio-scaled attribute is numeric attribute with the chrematistic that a value as being a multiple (or ratio) of another value. For example, you are 100 times richer with 100*thanwith*1. Data scales and types are important since the type of clustering used often depends on the data scale and type.

4.3 Traditional Clustering Algorithms

As clustering is an important technology related to applications, researchers use different models to define clustering problem and propose different ways to deal with the models. According to facts, in this chapter, we category the clustering algorithms into Partitional methods, Hierarchical methods, Density-methods, Grid-based methods and Model-based methods.

4.3.1 Partitional methods

Partitional methods have the following definition:

Definition 1: Given a set of input data set $D = \{x_1, x_2, \cdots, x_N\}$, where $x_i \in R^d$, $i = 1, \dots N$. Partitional methods attempt to seek K partitions of D, $C = \{C_1, C_2, \cdots, C_K\}$, $(K \leq N)$, such that the quality measure function $Q(C) = \sum_{k=1}^{K} \sum_{x_i \in C_k, x_j \in C_k} dist(x_i, x_j)$ is minimized, where $dist()$ is the distance function between the data objects.

Partitional methods create a one-level partitioning of the data objects. If K is the desired number of clusters, then partitional methods find all K clusters at once. Drineas et al. have proved that this problem is NP-hard [13]. In order to deal with the clustering problem described in definition 1, a number of partitional methods have been proposed. However, in this chapter, we shall only describe two approaches: K-means and K-medoid. Both these partitional methods are based on the idea that a center point can represent a cluster. However, there have been differences about the definition of 'center': For K-means we use the notion of a centroid which is the mean or median object of a group of data objects. In this case, the centroid almost never corresponds to an actual data object. For K-medoid we use the notion of a medoid which is the most central data object of a group of objects. According to the definition of a medoid, it is required to be an actual data object.

4.3.1.1 K-means

- The Framework of K-means.
  The K-means clustering algorithm, a top-ten algorithm in data mining area, is a very simple and widely used method. We immediately begin with a description of the framework of K-means, and then discuss

the details of each step in the framework. Algorithm 4.1 gives the framework of K-means [40]. Algorithm 4.1 gives a common framework of K-means. In the following:

Algorithm 4.1: K-means

Input: Data set D, Cluster number K

Output: Clustering result C

---

(1) Select K objects as the initial centroids
(2) Assign all data objects to the closest centroid
(3) Recompute the centroid of each cluster
(4) Repeat steps 2 and 3 until the centroids don't change to generate the cluster result C
(5) Return the cluster result C

---

we will discuss the details of each step in the framework.

- Initialization
  The first step of K-means is the initialization that choosing the K proper initial centroids, in this chapter, we call them as the seed objects. Seed objects can be the first K objects or K objects chosen randomly from the data set. A set of K objects that are separated from each other can be obtained by taking the centroid of the data as the first seed object and selecting successive seed objects which are at least a certain distance from the seed object already chosen [37]. The initial clustering result is formed by assigning each data object to the closest seed object. Different selection of the K seed objects could be introduced to different clustering results. This phenomenon is called as initialization sensitivity problem. There are two common ways to deal with the problem: the first one is to perform multiple runs, each with a different set of randomly chosen initial centroids, and the second one is based on the application knowledge. We will discuss the initialization sensitivity problem in Section 5.

- Updating partition
  Steps 2 and 3 in the framework are the updating partition part of K-means. Partitions are updated by reassigning objects to the clusters in an attempt to reduce the value. McQueen [37] defined a K-means pass as an assignment of all data objects to the closest cluster centroid, while the term 'pass' refers to the process of examining the cluster label of every object. The centroid of the gaining cluster is to re-computer after each new assignment. Otherwise, Forgy's [37] re-computering the cluster centroid after all patterns have been examined. In K-means, steps 2 and 3 are iteratively run until the $Q(C)$ value cannot be improved.

- Time and Space Complexity
Since only the vectors are stored, the space requirements are basically $O(m*n)$, where m is the number of the data objects and n is the number of attributes. The time cost are $O(I*K*m*n)$, where I is the number of iterations required for convergence, K is the number of clusters and $I_{jj}m$, $K_{jj}m$. Thus, K-means is linear in m, the number of points and is efficient, as well as simple.

- Adjusting the cluster number K
The selection of the cluster number K is one of the biggest drawbacks of K-means. When performing K-means, it is important to run diagnostic checks for determining the number of clusters in the data set at first. However, for real application, people cannot know how many clusters embedded in the data set. Adjusting the cluster number K is an acceptable way for dealing the selection of the cluster number K problem. Some clustering algorithms adjust the cluster number K by creating new clusters or by merging existing clusters if certain conditions are met. In one of the popular partitional clustering algorithms called ISODATA [37], these conditions are determined from parameters by the user of the program, for example, if a cluster has too many objects, it will be split; two clusters are merged if their cluster centroids are sufficiently close. Algorithm 4.2 shows the framework of ISODATA.

4.3.1.2 K-medoid Clustering

The K-medoid algorithm is a clustering algorithm related to the K-means algorithm and the medoid shift algorithm. The objective of K-medoid clustering is to find a non-overlapping set of clusters such that each cluster has a most representative object, i.e., an object that is most centrally located with respect to some measure, such as distance. These representative objects are called medoids and a medoid can be defined as the object of a cluster, whose average dissimilarity to all the objects in the cluster is minimal, i.e., it is a most centrally located point in the cluster. The most common realization of K-medoids clustering is the Partitioning Around Medoids (PAM) algorithm and as shows in Algorithm 4.3. It is more robust to noise and outliers as compared to K-means because it minimizes a sum of pairwise dissimilarities instead of a sum of squared Euclidean distances. However, finding a better medoid requires trying all points that are currently not medoids and are computationally expensive, it costs $O(K(m - K)^2)$.

Algorithm 4.2: the framework of ISODATA

Input: Data set D, Cluster number K

Output: Clustering result C

(1) Objects are assigned to the closest centroid and cluster centroids are re-computed. Iteratively repeated until no objects change clusters
(2) Clusters with "two few" objects are discarded
(3) Clusters are merged or split
    (a) If there are "too few" clusters compared to the number desired, then clusters are split;
    (b) If there are "too many" clusters compared to the number desired, then clusters are merged.
    (c) Otherwise the cluster splitting and merging phases alternate: clusters are merged if their centroids are close, while clusters are split if it contains "too many" objects.

Algorithm 4.3: Partitioning Around Medoids (PAM)

Input: Data Set D, Cluster number K
Output: Clustering result C

(1) Initialize: randomly select K of the m data objects as the mediods;
(2) Associate each data object to the closest medoid;
(3) for each medoid k
    (a) for each non-medoid data object o;
        (i) swap k and o and compute the total cost of the configuration;
(4) Select the configuration with the lowest cost;
(5) repeat steps 2 to 4 until there is no change in the medoid.

4.3.1.3 CLARA

CLARA (Clustering LARge Applications) is an adaptation of PAM for handling larger data sets, which was designed by Kaufman and Rousseeum in 1990. Instead of finding representative objects for the whole data set, CLARA, firstly, draws a sample of the data set by using sampling method; and then, applies PAM on the sample to find the medoids of the sample. The point is that, if the sample is drawn in a sufficiently random way, the medoids of the sample would approximate the medoids of the entire data set. To come up with better approximations, CLARA draws multiple samples and gives the best clustering as the output. In the accuracy case, the quality of a clustering is measured based on the average dissimilarity of all objects in the entire data set, and not only of the objects in the samples. Experiments results which reported in [46] indicate that five samples of size 40+2K give satisfactory results. Algorithm 4.4 shows the framework of CLARA algorithm. CLARA performs satisfactorily for large data sets by using PAM algorithm. Recall from Section 4.3.2 that each iteration of PAM is of $O(K(m - K)^2)$. But, for CLARA, by applying PAM just to the samples, each iteration is of $O(K(40+K)^2 + K(m-K))$. This explains why CLARA is more efficient than PAM for large values of m.

Algorithm 4.4: CLARA

Input: Data set D, Cluster number K;

Output: Clustering result C;

---

(1) $i=1$;
(2) while $i<5$
    (a) Draw a sample of $40 + 2K$ objects randomly from the entire data set, and call PAM algorithm to find K medoids of the sample.
    (b) For each object Oj in the entire data set, determine which of the K medoids is the most similar to Oj;
    (c) Calculate the average dissimilarity of the clustering obtained in the previous step. If this value is less than the current minimum, use this value as the current minimum, and retain the K medoids found in step 2 as the best set of medoids obtained so far;
(3) $i-i+1$;
(4) Return clustering result C

4.3.1.4 CLARANS

CLARANS uses a randomized search approach to improve on both CLARA and PAM. Algorithm 4.5 gives the conceptual description of CLARANS. From algorithm 4.5, we can see the difference between CLARANS and PAM: For a given current configuration, CLARANS algorithm does not consider all possible swaps of medoid and non-medoid objects, but rather, only a random selection of them and only until it finds a better configuration. Also, we can see the difference between CLARANS and CLARA: CLARANS works with all the data objects, however, CLARA only works with part of the entire data set. SD(CLARANS) and NSD(CLARAN) are two spatial data mining tools which contain CLARANS as a base algorithm. These tools added some capabilities related to cluster analysis.

Algorithm 4.5: CLARANS

Input: Data set D, Cluster number K;

Output: Clustering result C.

---

(1) Randomly select K candidate medoids;
(2) Randomly consider a swap of one of the selected medoid for a non-selected object;
(3) If the new configuration is better, then repeat step 2 with a new configuration;
(4) Otherwise, repeat step 2 with the current configuration unless a parameterized limit has been exceeded;
(5) Compare the current solution with any previous solutions and keep track of the best;
(6) Return to step 1 unless a parameterized limit has been exceeded;
(7) Return to the clustering result C

4.3.2 Hierarchical Methods

In hierarchical clustering, the goal is to produce a hierarchical series of nested clusters, ranging from clusters of individual points at the bottom to an all-inclusive cluster at the top. A diagram is used to graphically represent this hierarchy and is an inverted tree that describes the order in which objects are merged or clusters are split. The mathematical structure which a hierarchical clustering imposes on data is described as following. Give m objects which need to be clustered are denoted by the set $D=x_1,...,x_m$, where x_i is the i-th object. A partition C of D breaks D into subsets $C_1,C_2,...,C_K$ satisfying the following: $C_i \cap C_j = \emptyset$, $i, j = 1, ...K$, $i \neq j$ and $C_1 \cup C_2...C_K = D$, where $\cap$ stands for set intersection, $\cup$ stands for set union, and $\emptyset$ is the empty set. A clustering is a partition and the components of the partition are called clusters. Partition C is nested into partition C' if every component of C is a subset of a component of C'. That is, C' is formed by merging components of C.

For example, if the clustering C' with three clusters and clustering C with five clusters are defined as follows, then C is nested into C'. Both C and C' are the clustering of the set of objects $x1,x2,...,x10$ $C=(x1,x3)$, $(x5,x7)$, $(x2)$, $(x4,x6,x8)$, $(x9,x10)$ $C'=(x1,x3,x5,x7)$, $(x2,x4,x6,x8)$, $(x9,x10)$. Clusters $(x1,x3)$, $(x5,x7)$ in C are merged into cluster $(x1,x3,x5,x7)$ in C'. In the same way, clusters $(x2)$, $(x4,x6,x8)$ in C are merged into cluster $(x2,x4,x6,x8)$ in C'. However, cluster $(x9,x10)$ in C does not merge with any other clusters. But for partition $C''=(x1,x2,x3,x4)$, $(x5,x6,x7,x8)$, $(x9,x10)$, neither C nor C' is nested into it.

4.3.2.1 Agglomerative and Divisive algorithm

There are two basic approaches to generating a hierarchical clustering: agglomerative and divisive algorithms. An agglomerative algorithm for hierarchical clustering:

(1) assigns each of the *m* objects into an individual cluster to form the first clustering result;
(2) merges two or more of these trivial clusters according to the similarity measure, thus nesting the trivial clustering into a second partition;
(3) iteratively run step 2 until a single cluster containing all *m* objects or some requirements are achieved. A divisive algorithm performs the task in the reverse order: Start with one all-inclusive cluster and, at each step, split a cluster until only singleton cluster of individual point remains.

   Figure 4.3.1 shows the steps of these two kinds of methods. From the diagram aspect, the agglomerative clustering algorithms are working in a bottom-up manner, on the contrary, the divisive clustering algorithms are working in a top-down manner.

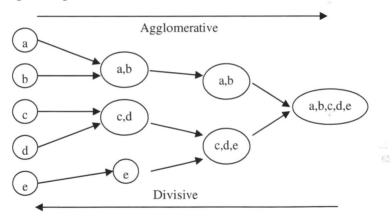

Figure 4.3.1: An example of Agglomeration and Division

4.3.2.2 Cluster Dissimilarity

In order to decide which clusters should be combined (for agglomerative), or where a cluster should be split (for divisive), a measure of dissimilarity between sets of observations is required. In most methods of hierarchical clustering, this is achieved by use of an appropriate metric (a measure of distance between pairs of observations), and a linkage criterion which specifies the dissimilarity of sets as a function of the pairwise distances of observations in the sets. The choice of an appropriate metric will influence the shape of the clusters, as some elements may be close to one another according to one distance and farther away according to another. For example, in a 2-dimensional space, the distance between the point (1,0) and the origin (0,0) is always 1 according to the usual norms, but the distance between the point (1,1) and the origin (0,0) can be 2, or 1 under Manhattan distance, Euclidean distance or maximum distance respectively. Some commonly used metrics for hierarchical clustering are described as follows:

- Euclidean distance: $||x_i - x_j||_2 = \sqrt{\sum_{l=1}^{n}(x_{il} - x_{jl})^2}$ $i, j = 1, ..., m$
- Squared Euclidean distance: $||x_i - x_j||_2^2 = \sum_{l=1}^{n}(x_{il} - x_{jl})^2$ $i, j = 1, ...,m$
- Manhattan distance: $||x_i - x_j||_1 = \sum_{l=1}^{n}|x_{il} - x_{jl}|$ $i, j = 1, ..., m$
- Maximum distance: $||x_i - x_j||_\infty = \max_l |x_{il} - x_{jl}|$ $l = 1, ..., n$ $i, j = 1, ...,m$
- Cosine similarity: $\cos(x_i, x_j) = \dfrac{x_i.x_j}{||x_i||||x_j||}$ $i, j = 1, ...,m$.

4.3.2.3 Divisive Clustering Algorithms

- Minimum Spanning Tree
  Let us introduce Minimum Spanning Tree, a simple divisive algorithm. This approach is the modified version of the single-link agglomerative method. The following framework shows the main steps of Minimum Spanning Tree:

Algorithm 4.6: Minimum Spanning Tree

---

(1) Compute a minimum spanning tree for the proximity graph;
(2) Create a new cluster by breaking the link corresponding to the smallest similarity;
(3) Repeat step 2 until only singleton cluster remains or some requirements achieved.

- Bi-Section-K-means
  Bi-Section-K-means is a variant of K-means. It is a good and fast divisive clustering algorithm. It frequently outperforms standard K-means as well as agglomerative clustering techniques. Bi-Section-K-means is defined as an outer loop around standard K-means. In order to generate K clusters, Bi-section-K-means repeatedly applies K-means. Bi-Section-K-means is initiated with the universal cluster containing all objects. Then it loops. It selects the cluster with the largest dissimilarity and it calls K-means in order to split this cluster into exactly two sub-clusters. The loop is repeated K-1 times such that K non-overlapping sub-

clusters are generated. Further, as Bi-Section-K-means is a randomized algorithm, we produce ten runs and average the obtained results.

Traceability. Concerning traceability, bi-section-K-means shares the problem that similarities in high-dimensional space are difficult to understand. In contrast to agglomerative algorithms, Bi-Section-K-means may incur that the two most similar terms are still split into different clusters, as a wrong decision at the upper level of generalization may jeopardize intuitive clusterings at the lower level.

Efficiency. The time complexity of Bi-Section-K-means the algorithm is $O(mK)$ where m is the number of objects and K is the number of clusters.

- DIANA

  DIANA is a hierarchical clustering technique and it works as follows: At first, there is one large cluster consisting of all m objects. And then, at each subsequent step, the largest available cluster is split into two clusters until finally all clusters comprise single objects. Thus, the hierarchy is built in n–1 steps. In the first step of an agglomerative method, all possible fusions of two objects are considered leading to $\frac{m(m-1)}{2}$ combinations. In the divisive method based on the same principle, there are possibilities to split the data into two clusters. This number is considerably larger than that in the case of an agglomerative method. To avoid considering all possibilities, the algorithm proceeds as follows:

(1) Find the object, which has the highest average dissimilarity to all other objects. This object initiates a new cluster—a sort of a splinter group;
(2) For each object i outside the splinter group compute;
(3) D_i = [average $d(i,j)$ $j \notin R_{\text{splinter group}}$]-[average $d(i,j)$ $j \in R_{\text{splinter group}}$]
(4) Find an object h for which the difference D_h is the largest. If D_h is positive, then h is, on the average close to the splinter group.
(5) Repeat Steps 2 and 3 until all differences D_h are negative. The data set is then split into two clusters.
(6) Select the cluster with the largest diameter. The diameter of a cluster is the largest dissimilarity between any two of its objects. Then divide this cluster, following steps 1–4.
(7) Repeat Step 5 until all clusters contain only a single object.

4.3.2.4 Agglomerative Clustering Algorithms

Given a set of m items to be clustered, and an $m \times m$ distance (or similarity) matrix, the basic process of agglomerative hierarchical clustering is as follows:

(1) Start by assigning each item to its own cluster, so that if you have *m* items, you now have *m* clusters, each containing just one item. Let the distances (similarities) between the clusters equal the distances (similarities) between the items they contain.

(2) Find the closest (most similar) pair of clusters and merge them into a single cluster, so that now you have one less cluster.

(3) Compute distances (similarities) between the new cluster and each of the old clusters.

(4) Repeat steps 2 and 3 until all items are clustered into a single cluster of size N. Thus, the agglomerative clustering algorithm will result in a binary cluster tree with single article clusters as its leaf nodes and a root node containing all the articles.

- Single Link
  For the single link or MIN version of hierarchical clustering, the proximity of two clusters is defined as minimum of the distance between any two objects in the different clusters. Single link is good at handling non-elliptical shapes, but is sensitive to noise and outliers. Figure 4.3.2(a) gives a sample similarity matrix for five objects and the dendrogram shows the series of merges derived by using the single link clustering method [Fig. 4.3.2(b)].

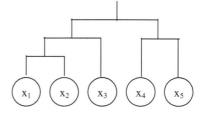

	x_1	x_2	x_3	x_4	x_5
x_1	1	0.92	0.09	0.65	0.21
x_2	0.92	1	0.72	0.61	0.50
x_3	0.09	0.72	1	0.45	0.30
x_4	0.65	0.61	0.45	1.	0.84
x_5	0.21	0.50	0.30	0.84	1

(a) similarity matrix for five objects   (b) Dendogram of single link

Figure 4.3.2: An example of Single Link

- Complete Link Clustering
  For the complete link version of hierarchical clustering, the proximity of two clusters is defined to be maximum of the distance (minimum of the similarity) between any two points in the different clusters. (The technique is called complete link because, if you start with all points as singleton clusters, and add links between points, strongest links first, then a group of points is not a cluster until all the points in it are completely linked.) Complete link is less susceptible to noise and outliers, but can break large clusters, and has trouble with convex shapes. The following table gives a sample similarity matrix and the

dendrogram shows the series of merges that result from using the complete link technique. Figure 4.3.3(a) gives a sample similarity matrix for five objects and the dendrogram shows the series of merges derived by using the single link clustering method [Fig. 4.3.3(b)].

	x_1	x_2	x_3	x_4	x_5
x_1	1	0.92	0.09	0.65	0.21
x_2	0.92	1	0.72	0.61	0.50
x_3	0.09	0.72	1	0.45	0.30
x_4	0.65	0.61	0.45	1.	0.84
x_5	0.21	0.50	0.30	0.84	1

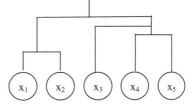

(a) Similarity matrix for five objects   (b) Dendogram of single link

Figure 4.3.3: An example of Complete Link Clustering

- **Average Link Clustering**
  For the group average version of hierarchical clustering, the proximity of two clusters is defined to be the average of the pairwise proximities between all pairs of points in the different clusters. Notice that this is an intermediate approach between MIN and MAX. This is expressed by the following equation: similarity (*cluster1*, *cluster2*) =
  $$\frac{\sum_{p1 \in cluster1, p2 \in cluster2} \text{similarity}(p1, p2)}{\| cluster1 \| * \| cluster2 \|}.$$
  Figure 4.3.4 gives a sample similarity matrix and the dendrogram shows the series of merges that result from using the group average approach. The hierarchical clustering in this simple case is the same as produced by MIN.

- **Ward's method**
  Ward's method says that the distance between two clusters, and , is how much the sum of squares will increase when we merge them:
  $$O(C_{k'}C_k) = \sum_{x_i \in Ck \cup Ck'} \| x_i - centroid_{C_k \cup C_{k'}} \|^2 - \sum_{xi \in Ck} \| x_i - centroid_{C_k} \|^2 - \sum_{xi \in Ck'} \| x_i - centroid_{C_{k'}} \|^2$$
  $$= \frac{num_{C_k} num_{C_{k'}}}{num_{C_k} + num_{C_{k'}}} \| centroid_{C_k} - centroid_{C_{k'}} \|^2$$

where $centroid_k$ is the center of cluster k, and num_k is the number of objects in it. O is called the merging cost of combining the clusters C_k and C'_k. With hierarchical clustering, the sum of squares starts out at zero (because every point is in its own cluster) and then grows as we merge clusters. Ward's method keeps this growth as small as possible. This is nice if you believe that the sum of squares should be small. Notice that the number of points shows up in O, as well as their geometric separation. Given two pairs of clusters whose centers are equally far apart, Ward's method will prefer to merge the smaller ones. Ward's method is both greedy, and constrained by previous choices as to which clusters to form. This means its sum-of-squares for a given number k of clusters is usually larger than the minimum for that k, and even larger than what k-means will achieve. If this is bothersome for your application, one common trick is use hierarchical clustering to pick K (see below), and then run k-means starting from the clusters found by Ward's method to reduce the sum of squares from a good starting point.

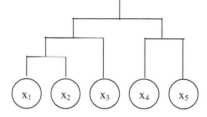

	x_1	x_2	x_3	x_4	x_5
x_1	1	0.92	0.09	0.65	0.21
x_2	0.92	1	0.72	0.61	0.50
x_3	0.09	0.72	1	0.45	0.30
x_4	0.65	0.61	0.45	1.	0.84
x_5	0.21	0.50	0.30	0.84	1

(a) Similarity matrix for five objects   (b) Dendrogram of average link

Figure 4.3.4: An example of Average Link Clustering

4.3.3 Density-based methods

Density-based approaches apply a local cluster criterion. Clusters are regarded as regions in the data space in which the objects are dense, and which are separated by regions of low object density (noise). These regions may have an arbitrary shape and the points inside a region may be arbitrarily distributed.

4.3.3.1 DBSCAN

The most popular density based clustering method is DBSCAN [22]. In contrast to many newer methods, it features a well-defined cluster model called "density-reachability". Similar to link-based clustering, it is based on connecting points within certain distance thresholds. DBSCAN has been applied to a 5-dimensional feature space created from several satellite images covering the area of California (5 different spectral channels: 1 visible, 2 reflected infrared, and 2 emitted (thermal) infrared). The images are taken from the roster data of the SEQUOIA 2000 Storage Benchmark. This kind of clustering application is one of the basic methods for automatic landuse detection from remote sensing data. The main idea of DBSCAN could be described as Fig. 4.3.5. From Fig. 4.3.5, we can find that DBSCAN's definition of a cluster is based on the notion of density reachability. Basically,

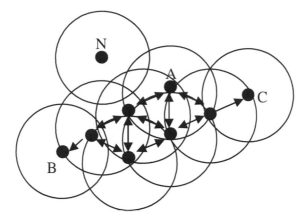

Figure 4.3.5: The basic idea of DBSCAN.

a point *q* is directly density-reachable from a point *p* if it is not farther away than a given distance (i.e., is part of its neighborhood) and if *p* is surrounded by sufficiently many points such that one may consider *p* and *q* to be part of a cluster. *q* is called density-reachable (note the distinction from "directly density-reachable") from *p* if there is a sequence *p1,...,pm* of points with *p1=p* and *pm =q* where each *pi*+1 is directly density-reachable from *pi*. Note that the relation of density-reachable is not symmetric. *q* might lie on the edge of a cluster, having insufficiently many neighbors to count as dense itself. This would halt the process of finding a path that stops with the first non-dense point. By contrast, starting the process with *q* would lead to *p* (though the process would halt there, *p* being the first non-dense point). Due to this asymmetry, the notion of density-connected is introduced: two points *p* and *q* are density-connected if there is a point *o* such that both *p* and *q* are density-reachable from *o*. Density-connectedness is symmetric. Algorithm 4.7 gives the main steps of DBSCAN. DBSCAN visits each point of the database, possibly multiple times (e.g., as candidates to different clusters). For practical considerations, however, the time complexity is mostly governed by the number of region Query invocations. DBSCAN executes exactly one such query for each point, and if an indexing structure is used that executes such a neighborhood query in $O(\log n)$, an overall runtime complexity of $O(v * log n)$ is obtained. Without the use of an accelerating index structure, the run time complexity is $O(n^2)$. Often the distance matrix of size $(n^2 - n)/2$ is materialized to avoid distance re-computations. This however also needs $O(n^2)$ memory. However, it only connects points that satisfy a density criterion in the original variant defined as a minimum number of other objects within this radius. A cluster consists of all density-connected objects (which can form a cluster of an arbitrary shape, in contrast to many other methods) plus all objects that are within these objects' range.

Another interesting property of DBSCAN is that its complexity is fairly low—it requires a linear number of range queries on the database—and that it will discover essentially the same results (it is deterministic for core and noise points, but not for border points) in each run, therefore there is no need to run it multiple times.

Algorithm 4.7: DBSCAN

---

(1)   select a point p

(2)   Retrieve all points density-reachable from p wrt ε and MinPts.

(3)   If p is a core point, a cluster is formed.

(4)   If p is a border point, no points are density-reachable from p and DBSCAN visits the next point of the database.

(5)   Continue the process until all of the points have been processed.

(6)   Return clustering results.

OPTICS [11] is a generalization of DBSCAN that removes the need to choose an appropriate value for the range parameter, and produces a hierarchical result related to that of linkage clustering. DeLi-Clu [5], Density-Link-Clustering combines ideas from single-link clustering and OPTICS, eliminating the parameter entirely and offering performance improvements over OPTICS by using an R-tree index. The key drawback of DBSCAN and OPTICS is that they expect some kind of density drop to detect cluster borders. Moreover they cannot detect intrinsic cluster structures which are prevalent in the majority of real life data. On data sets with, for example, overlapping Gaussian distributions—a common use case in artificial data—the cluster borders produced by these algorithms will often look arbitrary, because the cluster density decreases continuously. On a data set consisting of mixtures of Gaussians, these algorithms are nearly always outperformed by methods such as EM clustering that are able to precisely model this kind of data.

4.3.3.2 DENCLUE

DENCLUE (DENsity CLUstEring) [32] is a density clustering approach that takes a more formal approach to density based clustering by modeling the overall density of a set of points as the sum of "influence" functions associated with each point. The resulting overall density function will have local peaks, i.e., local density maxima, and these local peaks can be used to define clusters in a straightforward way. Specifically, for each data point, a hill climbing procedure finds the nearest peak associated with that point, and the set of all data points associated with a particular peak (called a local density attractor) becomes a (center-defined) cluster. However, if the density at a local peak is too low, then the points in the associated cluster are classified as noise and discarded. Also, if a local peak can be connected to a second local peak by a path of data points, and the density at each point on the path is above a minimum density threshold, then the clusters associated with these local peaks are merged. Thus, clusters of any shape can be discovered. DENCLUE is based on a well-developed area of statistics and pattern recognition which is known as "kernel density estimation" [18]. The goal of kernel density estimation (and many other statistical techniques as well) is to describe the distribution of the data by a function. For kernel density estimation, the contribution of each point to the overall density function is expressed by an "influence" (kernel) function. The overall density is then merely the sum of the influence functions associated with each point. The DENCLUE algorithm has two steps: a preprocessing step and a clustering step. In the pre-clustering step, a grid for the data is created by dividing the minimal bounding hyper-rectangle into d-dimensional hyper-rectangles with edge length 2σ. The rectangles that contain points are then determined. (Actually, only the occupied hyper-rectangles are constructed.) The hyper-rectangles are numbered with respect to a particular origin (at one edge of the bounding hyper-rectangle and these keys are stored in a search tree to provide efficient access in later processing. For each stored cell, the number of points, the sum of the points in the cell, and connections to neighboring population cubes are also stored. DENCLUE can be parameterized so that it behaves much like DBSCAN, but is much more efficient that DBSCAN. DENCLUE can also behave like K-means by choosing σ appropriately and by omitting the step that merges center-defined clusters into arbitrary shaped clusters. Finally, by doing repeated clusterings for different values of σ, a hierarchical clustering can be obtained.

4.3.4 Grid-based Methods

Grid-based clustering methods have been used in some data mining tasks of very large databases [29]. In the grid-based clustering, the feature space is divided into a finite number of rectangular cells, which form a grid. In this grid structure, all the clustering operations are performed. The grid can be formed in multiple resolutions by changing the size of the rectangular cells. Figure 4.3.6 presents a simple example of a hierarchical grid structure in three levels that is applied to a two-dimensional feature space. In the case of d-dimensional space, hyper rectangles (rectangular shaped cube [50]) of d-dimensions correspond to the cells. In the hierarchical grid structure, the cell size in the grid can be decreased in order to achieve a more precise cell structure. As in Fig. 4.3.6, the hierarchical structure can be divided into several levels of resolution. Each cell at the high level k is partitioned to form a number of cells at the next lower level $k+1$. The cells at the level $k+1$

are formed by splitting the cell at level k into smaller subcells. In the case of Fig. 1, each cell produces four subcells at the next lower level.

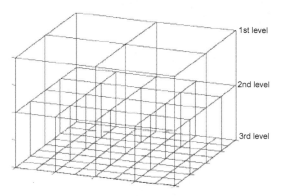

Figure 4.3.6: An example of a Grid-based structure

Grid-based clustering methods make it possible to form arbitrarily shaped, distance independent clusters. In these methods, the feature space is quantized into cells using a grid structure. The cells can be merged together to form clusters. Grid-based clustering was originally based on the idea of Warnekar and Krishna [58] to organize the feature space containing patterns. Erich [50] has used topological neighbor search algorithm to combine the grid cells to form clusters. Agrawal et al. [1] have presented a density-based clustering method using grid which named CLIQUE. In this chapter, we will discuss STING and WaveCluster algorithms, and CLIQUE will be detailed in the section "High dimensional clustering algorithm".

4.3.4.1 STING

A new statistical information grid-based method (STING) was proposed in [57] to efficiently process many common "region oriented" queries on a set of points. Region oriented queries are defined later more precisely but informally, they ask for the selection of regions satisfying certain conditions on density, total area, etc. Algorithm 4.8 shows the framework of STING.

In the above algorithm, Step 1 takes constant time. Steps 2 and 3 require a constant time for each cell to calculate the confidence interval or estimate proportion range and also a constant time to label the cell as relevant or not relevant. This means that we need constant time to process each cell in Steps 2 and 3. The total time is less than or equal to the total number of cells in our hierarchical structure. Notice that the total number of cells is 1.33K, where K is the number of cells at bottom layer. We obtain the factor 1.33 because the number of cells of a layer is always one-fourth of the

---

Algorithm 4.8: STING

Input: Data set D;

Output: Clustering results

---

1. Determine a layer to begin with.
2. For each cell of this layer, we calculate the confidence interval (or estimated range) of probability that this cell is relevant to the query.
3. From the interval calculated above, we label the cell as *relevant* or *not relevant*.
4. If this layer is the bottom layer, go to Step 6; otherwise, go to Step 5.
5. We go down the hierarchy structure by one level. Go to Step 2 for those cells that form the *relevant* cells of the higher level layer.
6. If the specification of the query is met, go to Step 8; otherwise, go to Step 7.
7. Retrieve those data that fall into the *relevant* cells and do further processing. Return the result that meets the requirement of the query. Go to Step 9.
8. Find the regions of *relevant* cells. Return those regions that meet the requirement of the query. Go to Step 9.
9. Stop and return clustering results.

---

number of cells of the layer one level lower. So the overall computation complexity on the grid hierarchy structure is $O(K)$. Usually, the number of cells needed to be examined is much less, especially when many cells at high layers are not relevant. In Step 8, the time it takes to form the regions is linearly proportional to the number of cells. The reason is that for a given cell, the number of cells need to be examined is constant because both the specified density and the granularity can be regarded as constants during the execution of a query and in turn the distance is also a constant since it is determined by the specified density. Since we assume each cell at bottom layer usually has several dozens to several thousands objects, $K \ll N$. So, the total complexity is still $O(K)$. Usually, we do not need to do Step 7 and the overall computational complexity is $O(K)$. In the extreme case that we need to go to Step 7, we still do not need to retrieve all data from database. Therefore, the time required in this step is still less than linear. So, this algorithm outperforms other approaches greatly.

WaveCluster [52] is a clustering technique that interprets the original data as a two-dimensional signal and then applies signal processing techniques (the wavelet transform) to map the original data to a new space where cluster identification is more straightforward. More specifically, WaveCluster defines a uniform two-dimensional grid on the data and represents the points in each grid cell by the number of points. Thus, a collection of two-dimensional data points becomes an image, i.e., a set of "gray-scale" pixels, and the problem of finding clusters becomes one of image segmentation. While there are a number of techniques for image segmentation, wavelets have a couple of features that make them an attractive choice. First, the wavelet approach naturally allows for a multiscale analysis, i.e., the wavelet transform allows features, and hence, clusters, to be detected at different scales, e.g., fine, medium, and coarse. Secondly, the wavelet transform naturally lends itself to noise elimination. Algorithm 4.9 shows the framework of WaveCluster. In summary, the key features of WaveCluster are order independence, there is no need to specify a number of clusters (although it is helpful to know this in order to figure out the right scale to look at), speed (linear), the elimination of noise and outliers, and the ability to find arbitrarily shaped clusters. While the WaveCluster approach can theoretically be extended to more than two dimensions, it seems unlikely that WaveCluster will work well (efficiently and effectively) for medium or high dimensions.

Algorithm 4.9: WaveCluster

Input: Data set D

Output: Clustering results

---

(1) Create a grid and assign each data object to a cell in the grid. The grid is uniform, but the grid size will vary for different scales of analysis. Each grid cell keeps track of the statistical properties of the points in that cell, but for wave clustering only the number of points in the cell is used.

(2) Transform the data to a new space by applying the wavelet transform. This results in 4 "subimages" at several different levels of resolutionan "average" image, an image that emphasizes the horizontal features, an image that emphasizes vertical features and an image that emphasizes corners.

(3) Find the connected components in the transformed space. The average subimage is used to find connected clusters, which are just groups of connected "pixels," i.e., pixels which are connected to one another horizontally, vertically, or diagonally.

(4) Map the clusters labels of points in the transformed space back to points in the original space. WaveCluster creates a lookup table that associates each point in the original with a point in the transformed space. Assignment of cluster labels to the original points is then straightforward.

4.3.5 Model-based Methods

It was realized early on that cluster analysis can be based on probabilistic or stochastic models. This realization has provided insight into when a particular clustering method can be expected to work well (i.e., when the data conform to the model) and has led to the development of new clustering methods. It has also been shown that some of the most popular heuristic clustering methods, such as the k-means algorithm, are approximate estimation methods for particular probabilistic models [26]. Finite mixture models are a flexible and powerful probabilistic modeling tool for univariate and multivariate data. It is assumed that the data are generated by a mixture of underlying probability distributions for multiple components where each component represents a different group or cluster.

Given data D with independent multivariate observations $x_1, x_2, \ldots, x_m$, the joint likelihood of D is:

$$P(D \mid \Theta) = \prod_{i=1}^{m} P(x_i \mid \Theta) = \prod_{i=1}^{m} \sum_{k=1}^{K} \tau_i P(x_i \mid \theta_k)$$

Where $\Theta = \{\theta_1, \theta_2, \ldots, \theta_K, \tau_1, \tau_2, \ldots, \tau_K\}$ represents the set of all model parameters for the mixture model. τ_K is the probability that an observation belongs to the kth component ($\tau_K > 0, \sum_{k=1}^{K} \tau_K = 1$). Model parameter learning amounts to finding the maximum a posteriori (MAP) parameter estimate, given the data set D, i.e., $\widehat{\Theta} = \arg\max[P(D \mid \Theta)P(\Theta)]$. If we take a noninformative prior on Θ, learning degenerates to maximum likelihood estimation (MLE), i.e., $\widehat{\Theta} = \operatorname{argmax} P(D \mid \Theta)$.

4.3.5.1 EM Algorithm

The expectation-maximization (EM) algorithm [20] is a general approach to MLE problems in which the data can be viewed as consisting of m multivariate observations (x_i, z_i), where x_i is observed object but z_i is unobserved.

(1) Basic EM If (x_i, z_i) are independent and identically distributed according to a probability distribution P with parameter θ, then the complete-data likelihood is:

$$L_C(D, Z \mid \theta) = \prod_{i=1}^{m} P(x_i, z_i \mid \theta)$$

The observed data likelihood can be obtained by integrating out Z from the complete-data likelihood, $L_0(D \mid \theta) = \int L_C(D, Z \mid \theta) dZ$. The MLE for θ based on the observed data maximizes $L_0(D \mid \theta)$ with respect to θ. Without knowing the missing data, the EM algorithm alternates between two steps. In the E-step, the conditional expectation of the complete-data log-likelihood, given the observed data and the current parameter estimates is computed. In the M-step, the parameters that maximize the expected log-likelihood from E-step are determined.

(2) EM algorithm for Mixture models. In the EM algorithm for mixture models, Equation $P(D \mid \Theta) = \prod_{i=1}^{m} P(x_i \mid \Theta) = \prod_{i=1}^{m} \sum_{k=1}^{K} \tau_i P(x_i \mid \theta_k)$ can be rewritten as a log-likelihood function of parameter for given observed data D: $\ell(\Theta; D) = \sum_{i=1}^{m} \log[\sum_{k=1}^{K} \tau_k P(x_i \mid \theta_k)]$. The "complete-data" are considered to be (x_i, z_i), where $z_i = (z_{i1}, \ldots, z_{iK})$ is the unobserved binary K-dimensional vectors such that $z_{ik} = 1$ if and only if x_i arises from the k-th component[60].

Assuming that each z_i is independent and identically distributed according to multinomial distribution with probabilities $\tau_1, \tau_2, ..., \tau_K$ for the K clusters, and the density of an observation x_i given z_i is given by $\prod_{k=1}^{K} P(x_i \mid \theta_k)^{z_{ik}}$, the resulting complete-data log-likelihood is $\ell_c(\Theta; D) = \sum_{i=1}^{m} \sum_{k=1}^{K} z_{ik} \log[\tau_k P(x_i \mid \theta_k)]$, which is also known as the classification log-likelihood in the clustering literature. The EM algorithm can then be viewed as operating in two steps. In the E-step, we calculate the class-conditional probability $P(z_i \mid x_i, \Theta)$ for each sequence under each of the K clusters using the current parameter. In the M-step, we update by weighing each sequence according to its class-conditional probability. Thus the EM algorithm is guaranteed to lead to a sequence of Θ's which have non-decreasing likelihood, i.e., under fairly broad conditions it will find at least a local maximum. Algorithm 4.10 shows the main steps of EM algorithm and The implement of EM clustering was embedded in MCLUST R package (http://www.stat.washington. edu/mclust/).

Algorithm 4.10: EM clustering

Input: Data set D

Output: Clustering result

---

(1) First, initialize the parameters θ to some random values.
(2) Compute the best value for Z given these parameter values.
(3) Then, use the just-computed values of Z to compute a better estimate for the parameters θ. Parameters associated with a particular value of Z will use only those data points whose associated latent variable has that value.
(4) iterate steps 2 and 3 until convergence.
(5) Return clustering result.

4.3.5.2 Extensions of EM Algorithm

The EM algorithm for clustering has a number of limitations. First, the rate of convergence can be very slow. This does not appear to be a problem in practice for well-separated mixtures when started with reasonable values. Second, the number of conditional probabilities associated with each observation is equal to the number of components in the mixture, so that the EM algorithm for clustering may not be practical for models with very large number of clusters. Finally, EM breaks down when the covariance matrix corresponding to one or more components becomes ill-conditioned (singular or nearly singular). If EM for a model with a certain number of components is applied to a mixture with fewer groups than the number of mixture components, then it may fail due to ill-conditioning. A number of variants of the EM algorithm have been proposed for model-based clustering, some of which can avoid the limitations of the EM algorithm. CEM and SEM [19] are two widely used variants. The classification EM (CEM) algorithm can be regarded as a classification version of the EM algorithm, where the complete log-likelihood is maximized. It incorporates a classification step (C-step) between the E-step and the M-step of the standard EM algorithm by using a MAP principle. In the C-step, each object x_i is assigned to the cluster which provides the maximum posterior probability. It has been shown that the k-means algorithm is exactly the CEM algorithm for a Gaussian mixture with equal proportions and a common covariance matrix of the form. Since most of the classical clustering criteria can be analyzed as classification maximum likelihood criteria, the CEM algorithm turns out to be quite a general clustering algorithm. However, from the practical point of view, the solution provided by the CEM algorithm does depend on its initial position, especially when the clusters are not well separated. Celeux et al. considered a stochastic version of EM as well as the CEM algorithm in the context of computing the MLE for finite mixture models. They call it the stochastic EM (SEM) algorithm. With the SEM algorithm, the current posterior probabilities are used in a stochastic E-step (Sstep), wherein each observation object xi is assigned to one of the K clusters according to the posterior probability distributions for all clusters. Numerical experiments have shown that SEM performs well and can overcome some of the limitations of the EM algorithm.

4.4 High-dimensional clustering algorithm

Data collected in the world are so large that it is becoming increasingly difficult for users to access them. Knowledge Discovery in Databases (KDD) is the non-trivial process of identifying valid, novel, potentially useful and ultimately understandable patterns in data [23]. The KDD process is interactive and iterative, involving numerous steps. Data mining is one such step in the KDD process. In this section, we focus on the high-dimensional clustering problem, which is one of the most useful tasks in data mining for discovering groups and identifying interesting distributions and patterns in the underlying data. Thus, the goal of clustering is to partition a data set into subgroups such that objects in each particular group are similar and objects in different groups are dissimilar [15]. According to [31], four problems need to be overcome for high-dimensional clustering:

(1) Multiple dimensions are hard to think in, impossible to visualize, and due to the exponential growth of the number of possible values with each dimension, complete enumeration of all subspaces becomes

intractable with increasing dimensionality. This problem is referred to as the curse of dimensionality.

(2) The concept of distance becomes less precise as the number of dimensions grows, since the distance between any two points in a given dataset converges. The discrimination of the nearest and farthest point in particular becomes meaningless: $\lim_{d \to \infty} \frac{dist_{max} - dist_{min}}{dist_{min}} \to 0$.

(3) A cluster is intended to group objects that are related, based on observations of their attribute's values. However, given a large number of attributes some of the attributes will usually not be meaningful for a given cluster. For example, in newborn screening, a cluster of samples might identify newborns that share similar blood values, which might lead to insights about the relevance of certain blood values for a disease. But for different diseases, different blood values might form a cluster, and other values might be uncorrelated. This is known as the local feature relevance problem: different clusters might be found in different subspaces, so a global filtering of attributes is not sufficient.

(4) Given a large number of attributes, it is likely that some attributes are correlated. Hence, clusters might exist in arbitrarily oriented affine subspaces. Recent research by Houle et al. [42] indicates that the discrimination problems only occur when there is a high number of irrelevant dimensions, and that shared-nearest-neighbor approaches can improve results.

Lots of clustering algorithms have been proposed to high dimensional clustering problem. In general, the algorithmic approaches for finding these subspaces (i.e., traversing the search space of all possible axis-parallel subspaces) can be divided into the following two categories: Bottom-up approaches and Top-down approaches. In the following sections, we will discuss these two different ways in detail.

4.4.1 Bottom-up Approaches

The exponential search space that needs to be traversed is equivalent to the search space of the frequent item set problem in market basket analysis in transaction databases [10]. Each attribute represents an item and each subspace cluster is a transaction of the items representing the attributes that span the corresponding subspace. Finding item sets with frequency 1 then relates to finding all combinations of attributes that constitute a subspace containing at least one cluster. This observation is the rationale of most bottom-up subspace clustering approaches. The subspaces that contain clusters are determined starting from all one-dimensional subspaces that accommodate at least one cluster by employing a search strategy similar to frequent item set mining algorithms. To apply any efficient frequent item set mining algorithm, the cluster criterion must implement a downward closure property (also called monotonicity property): If subspace S contains a cluster, then any subspace $T \supseteq S$ must also contain a cluster. The reverse implication, if a subspace T does not contain a cluster, then any super space $S \supseteq T$ also cannot contain a cluster, can be used for pruning, that is, excluding specific subspaces from consideration. Let us note that there are bottom-up algorithms that do not use an APRIORI-like subspace search, but instead apply other search heuristics. In another way, the bottom-up approaches are also called as subspace clustering.

CLIQUE [1], the pioneering approach to subspace clustering, uses a grid-based clustering notion. The data space is partitioned by an axis-parallel grid into equal units of width ζ. Only units which contain at least τ points are considered as dense. A cluster is defined as a maximal set of adjacent dense units. Since dense units satisfy the downward closure property, subspace clusters can be explored rather efficiently in a bottom-up way. Starting with all one-dimensional dense units, $(k+1)$-dimensional dense units are computed from the set of k-dimensional dense units in an APRIORI-like style. If a $(k+1)$-dimensional unit contains a projection onto a k-dimensional unit that is not dense, then the $(k+1)$-dimensional unit also cannot be dense. Further, a heuristic that is based on the minimum description length principle is introduced to discard candidate units within less interesting subspaces (i.e., subspaces that contain only a very small number of dense units). This way, the efficiency of the algorithm is enhanced but at the cost of incomplete results, namely some true clusters are lost. There are some variants of CLIQUE. The method ENCLUS [11] also relies on a fixed grid, but searches for subspaces that potentially contain one or more clusters rather than for dense units. Three quality criteria for subspaces are introduced, one implementing the downward closure property. The method MAFIA [45] uses an adaptive grid. The generation of subspace clusters is similar to CLIQUE. Another variant of CLIQUE, called nCluster [41], allows overlapping windows of length δ as one-dimensional units of the grid. In summary, all grid-based methods use a simple but rather efficient cluster model. The shape of each resulting cluster corresponds to a polygon with axis-parallel lines in the corresponding subspace. Obviously, the accuracy and efficiency of CLIQUE and its variants primarily depend on the granularity and the positioning of the grid. A higher grid granularity results in higher runtime requirements but will most likely produce more accurate results. SUBCLU [16] uses the DBSCAN cluster model of density connected sets. It is shown that density-connected sets satisfy the downward closure property. This enables SUBCLU to search for density based clusters in subspaces in an APRIORI-like style. The resulting clusters may exhibit an arbitrary shape and size in the corresponding subspaces. RIS [39] is a subspace ranking algorithm that uses a quality criterion to rate the interestingness of subspaces. This criterion is based on the monotonicity of core points which are the central concept of the density-based clustering notion of DBSCAN. An Apriori-like subspace generation method (similar to SUBCLU) is used to compute all relevant subspaces and rank them by interestingness. The clusters can be computed in the generated subspaces using any clustering method of choice. SURFING [14] is a subspace ranking algorithm that does not rely on a global density threshold. It computes the interestingness of a subspace based on the distribution of the k-nearest neighbors of all data points in the corresponding projection. An efficient, bottom-up subspace expansion heuristics ensures that less interesting subspaces are not generated for examination. More subspace clustering algorithms in detailed, please reference to [53, 56, 35].

4.4.2 Top-down Approaches

The rationale behind Top-down approaches is to determine the subspace of a cluster starting from the full-dimensional space. This is usually done by determining a subset of attributes for a given set of points (potential cluster members) such that the points meet the given cluster criterion when projected onto the corresponding subspace. Obviously, the dilemma is that for the determination of the subspace of a cluster, at least some cluster members must be identified. On the other hand, in order to determine cluster memberships, the subspace of each cluster must be known. To escape from this circular dependency, most top-down approaches rely on a rather strict assumption, which we call the *locality assumption*. It is assumed that the subspace of a cluster can be derived from the local neighborhood (in the full-dimensional data space) of the cluster center or the cluster members. In other words, it is assumed that even in the full-dimensional space, the subspace of each cluster can be learned from the local neighborhood of cluster representatives or cluster members. Other top-down approaches that do not rely on the locality assumption use random sampling in order to generate a set of potential cluster members. According to the top-down approaches working way, it is also called as projective clustering. Projective clustering is an efficient way of dealing with high dimensional clustering problems. Explicitly or implicitly, projective clustering algorithms assume the following definition: Give a data set D of n-dimensional data objects, a projected cluster is defined as a pair (C_k, S_k), where C_k is a subset of data objects and S_k is a subset of attributes such that the data objects in C_k are projected along each attribute in S_k onto a small range of values, compared to the range of values of the whole data set in S_k, and the data objects in C_k are uniformly distributed along every other attributes not in S_k. The task of projective clustering is to search and report all projective clusters in the search space.

PROCLUS [9] is one of the classical projective clustering algorithms. It discovers groups of data objects located closely in each of the related dimension in its associated subspace. In such case, the data objects would spread along certain directions which are parallel to the original data axes. ORCLUS [10] aims to detect arbitrarily oriented subspaces formed by any set of orthogonal vectors. EPCH [38] is focused on uncovering projective clusters with varying dimensionality, without requiring users to input the expected average dimensionality l of the associated subspace and the number of clusters K that inherently exists in the data set. The d-dimensional histogram created with equal width, is used to capture the dense units and their locations in the d-dimensional space. A compression structure is used to store these dense units and their locations. At last, a search method is used to merge similar and adjacent dense units and form subspace clusters. P3C [44] can effectively discover projective clusters in the data while minimizing the number of required parameters. P3C also does not need the number of projective clusters as input and can discover the true number of clusters. There are three steps consisted in P3C. Firstly, regions corresponding to the clusters on each attribute are discovered. Secondly, a cluster core structure described by a combination of the detected regions is designed to capture the dense areas in a high dimensional space. Thirdly, cluster cores are refined into projective clusters, outliers are identified, and the relevant attributes for each cluster are determined. STATPC [43] uses a varying width hyper-rectangle structure to find out the dense areas embedded in the high dimensional space. By using a spatial statistical method, all dense hyper-rectangles are found. A heuristic search process is run to merge these dense hyper-rectangles and clustering results are generated. The clusters of projective clustering are defined as the dense areas in corresponding subsets of attributes. In projective clustering, it is a common way that a hyper-rectangle structure is used to find out the dense areas in the d-dimensional space at first; and then, a search method is run to merge these hyper-rectangles for generating clusters. Because the dense area is captured by the hyper-rectangle structure, it is important to define the structure before clustering. There are two kinds of hyper-rectangle structures used in projective clustering—the equal width hyper-rectangle structure and the varying width hyper-rectangle structure. For the equal width hyper-rectangle structure, each dimension is divided into equal width intervals, and the hyper-rectangles are constructed by these intervals, for instance, the d-dimensional histogram is used as the first step in the construction of hyper-rectangle structure in EPCH.

4.4.3 Other Methods

Hybrid clustering algorithms do not belong to bottom-up or top-down approaches. Algorithms that do not aim at uniquely assigning each data point to a cluster nor at finding all clusters in all subspaces are called hybrid algorithms. Some hybrid algorithms offer the user an optional functionality of a pure projected clustering algorithm. Others aim at computing only the subspaces of potential interest rather than the final clusters. Usually, hybrid methods that report clusters allow overlapping clusters, but do not aim at computing all clusters in all subspaces. DOC [49] uses a global density threshold to define a subspace cluster by means of hypercubes of fixed side-length w containing at least α points. A random search algorithm is proposed to compute such subspace clusters from a starting seed of sampled points. A third parameter β specifies the balance between the number of points and the dimensionality of a cluster. This parameter affects the dimensionality of the resulting clusters, and thus DOC usually also has problems with subspace clusters of significantly different dimensionality. Due to the very simple clustering model, the clusters may contain additional noise points (if w is too large) or not all points that naturally belong to the cluster (if w is too small). One run of DOC may (with a certain probability) find one subspace cluster. If k clusters need to be identified, DOC has to be applied at least k times. If the points assigned to the clusters found so far are excluded from subsequent runs, DOC can be considered as a pure projected clustering algorithm because each point is uniquely assigned to one cluster or to noise (if not assigned to a cluster). On the other hand, if the cluster points are not excluded from subsequent runs, the resulting clusters of multiple runs may overlap. Usually, DOC cannot produce all clusters in all subspaces. MINECLUS [60] is based on a similar idea as DOC, but proposes a deterministic method to find an optimal projected cluster, given a sample seed point. The authors transform the problem into a frequent item set mining problem and employ a modified frequent pattern tree growth method. Further heuristics are introduced to enhance efficiency and accuracy.

DiSH [6] follows a similar idea as PreDeCon but uses a hierarchical clustering model. This way, hierarchies of subspace clusters can be discovered, that is, the information that a lower-dimensional cluster is embedded within a higher-dimensional one. The distance between points and clusters reflects the dimensionality of the subspace that is spanned by combining the corresponding subspace of each cluster. As in COSA, the weighting of attributes is learned for each object, not for entire clusters. The learning of weights, however, is based on single attributes, not on the entire feature space. DiSH uses an algorithm that is inspired by the density-based hierarchical clustering algorithm OPTICS. However, DiSH extends the cluster ordering computed by OPTICS in order to find hierarchies of subspace clusters with multiple inclusions (a lower-dimensional subspace cluster may be embedded in multiple higher-dimensional subspace clusters). SCHISM [51] mines interesting subspaces rather than subspace clusters, hence, it is not exactly a subspace clustering algorithm, but solves a related problem: finding subspaces to look for clusters. It employs a grid-like discretization of the database and applies a depthfirst search with backtracking to find maximally interesting subspaces. FIRES [47] computes one-dimensional clusters using any clustering technique the user is most accomplished with in a first step. These one-dimensional clusters are then merged by applying a "clustering of clusters." The similarity of clusters is defined by the number of intersecting points. The resulting clusters represent hyper-rectangular approximations of the true subspace clusters. In an optional postprocessing step, these approximations can be refined by again applying any clustering algorithm to the points included in the approximation projected onto the corresponding subspace. Though using a bottom-up search strategy, FIRES is rather efficient because it does not employ a worst-case exhaustive search procedure but a heuristic that is linear in the dimensionality of the data space. However, this performance boost is paid for by an expected loss of clustering accuracy. It cannot be specified whether the subspace clusters produced by FIRES may overlap or not. In general, the clusters may overlap, but usually FIRES cannot produce all clusters in all subspaces.

4.5 Constraint-based Clustering Algorithm

In computer science, constrained clustering is a class of semi-supervised learning algorithms [36]. Typically, constrained clustering incorporates either a set of *must-link constraints*, *cannot-link constraints*, or both, with a data clustering algorithm. Both a must-link and a cannot-link constraint define a relationship between two data instances. A must-link constraint is used to specify that the two instances in the must-link relation should be associated with the same cluster. A cannot-link constraint is used to specify that the two instances in the cannot-link relation should not be associated with the same cluster. These sets of constraints acts as a guide for which a constrained clustering algorithm will attempt to find clusters in a data set which satisfy the specified must-link and cannot-link constraints. Some constrained clustering algorithms will abort if no such clustering exists which satisfies the specified constraints. Others will try to minimize the amount of constraint violation should it be impossible to find a clustering which satisfies the constraints.

4.5.1 COP K-means

In the context of partitioning algorithms, instance level constraints are a useful way to express a prior knowledge about which instances should or should not be grouped together. Consequently, we consider two types of pair-wise constraints:

- Must-link constraints specify that two instances have to be in the same cluster.
- Cannot-link constraints specify that two instances must not be placed in the same cluster.

The must-link constraints define a transitive binary relation over the instances. Consequently, when making use of a set of constraints (of both kinds), we take a transitive closure over the constraints. The full set of derived constraints is then presented to the clustering algorithm. In general, constraints may be derived from partially labeled data or from background knowledge about the domain or data set.

Algorithm 4.11 gives the framework of COP K-means algorithm. The major modification is that, when updating cluster assignments, we ensure that none of the specified constraints are violated. We attempt to assign each point d_i to its closest cluster C_j. This will succeed unless a constraint would be violated. If there is another point $d=$ that must be assigned to the same cluster as d, but that is already in some other cluster, or there is another point d_F that cannot be grouped with d but is already in C, then d_i cannot be placed in C. We continue down the sorted list of clusters until we find one that can legally host d. Constraints are never broken; if a legal cluster cannot be found for d, the empty partition is returned. An interactive demo of this algorithm can be found at http://www.cs.cornell.edu/home/wkiri/cop-kmeans/.

4.5.2 MPCK-means

Given a set of data objects D, a set of must-link constraints M, a set of cannot-link constraints C, corresponding cost sets W and $\overline{W}$, and the desired number of clusters K, MPCK-Mmeans finds a disjoint K-partitioning $\{C_k\}_{k=1}^K$ of D (with each cluster having a centroid μ_k and a local weight matrix Ak) such that the objective function is (locally) minimized [17]. The algorithm integrates the use of constraints and metric learning. Constraints are utilized during cluster initialization and when assigning points to clusters, and the distance metric is adapted by re-estimating the weight matrices Ak during each iteration based on the current cluster assignments and constraint violations. Algorithm 4.12 gives the pseudocode of MPCK-means.

Algorithm 4.11: COP-K-means
Input: data set D, must-link constraints $Con_= \subseteq D \times D$, cannot-link constraints $Con \subseteq D \times D$
Output: Clustering result
(1) Let $C_1, ..., C_k$ be the initial cluster centers.   (2) For each point d_i in D, assign it to the closest cluster C_j such that violate-constraints (d_i, C_j, $Con_=$, Con) is false. If no such cluster exists, fail (return {}).   (3) For each cluster C_i, update its center by averaging all of the points d_j that have been assigned to it.   (4) Iterate between (2) and (3) until convergence.   (5) Return {$C_1, ..., C_k$}.   violate-constraints(data point d, cluster C, must-link constraints $Con \subseteq D \times D$, cannot-link constraints $Con_{\neq} \subseteq D \times D$)   (1) For each ($d,d_=$) $Con_=$: if $d_= \notin C$, return true.   (2) For each ($d,d_{\neq}$) $Con_{\neq}$: if $d_{\neq} \notin C$, return true.   (3) Otherwise, return false

4.5.3 AFCC

AFCC is based on an iterative reallocation that partitions a data set into an optimal number of clusters by locally minimizing the sum of intra-cluster distances while respecting as many as possible of the constraints provided. AFCC alternates between membership updating step and centroid estimation step while generating actively at each iteration new candidate pairs for constraints [28]. After the initialization step, we continue by computing α, the factor that will ensure a balanced influence from the constrained data and unlabeled patterns than β, the factor that will determine which term of the membership updating equation will dominate. Afterwards, memberships will be updated. In the second step, based on the cardinalities of different clusters, spurious clusters will be discarded, thus obtaining the centroids of good clusters. At this time, a data partition is available, AFCC will then try to identify least well defined cluster and selects in an active manner good candidates for the need of generating maximally informative constraints. As distance $d(x_i, \mu_j)$ between a data item x_i and a cluster centriod μ_j, one can use either the ordinary Euclidean distance when the clusters are assumed to be spherical or the Mahalanobis distance when they are assumed to be elliptical: $d^2(x_i, \mu_k) = |C_k|^{1/n}(x_i - \mu_k)^T \delta_k^{-1}(x_i - \mu_k)$, where n is the dimension of the space considered and C_k is the covariance matrices of the cluster k:

$$\delta_k = \frac{\sum_{i=1}^{M} u_{ik}^2 (x_i - \mu_k)(x_i - \mu_k)^T}{\sum_{i=1}^{M} u_{ik}^2}$$

Algorithm 4.12: MPCK-means

Input: Data set D, must-link set M, cannot-link set CA, cluster number K, constraint costs W and $\overline{W}$

Output: Clustering result $\{C_k\}_{k=1}^K$

---

(1) Initialize clusters:

    (1.1) create the λ neighborhoods $\{N_p\}_{p=1}^\lambda$ from M and CA

    (1.2) if $\lambda \geq K$

        Initialize $\{\mu^0_k\}_{k=1}^K$ using weighted farthest-first traversal starting from the largest N_p

        Else if $\lambda < K$

            Initialize $\{\mu^0_k\}_{k=1}^K$ with centroids of $\{N_p\}_{p=1}^\lambda$

            Initialize remaining clusters at random

(2) Repeat until convergence

    (2.1) assign_cluster: assign each data object x_i to cluster k.

    (2.2) estimate means: $\{\mu_k^{(t+1)}\}_{k-1}^K \leftarrow \{\dfrac{1}{|C_k^{(t+1)}|}\sum_{x \in C_k^{(t+1)}} x\}_{k=1}^K$

    (2.3) update_metrics: $A_k = |C_k|\,(\Sigma_{x_i \in C_k}(x_i - \mu_k)(x_i - \mu_k)^T)$

$+ \Sigma_{(x_i, x_j) \in M_k} \dfrac{1}{2} w_{ij}(x_i - x_j)(x_i - x_j)^T + \Sigma_{(x_i, x_j) \in CA_k} \overline{w}_{ij}(x_i - x_j)(x_i - x_j)^T)^{-1}$

    (2.4) $t \leftarrow t + 1$

(3) Return Clustering result.

When the Mahalanobis distance is employed, the computation of δ_k are performed at the beginning of the main loop, right before the update of β. The AFCC algorithm runs in $O(MK^2p)$ time, where M is the number of data objects, K is the number of clusters, and p is the dimension of the data points.

4.6 Consensus Clustering Algorithm

Consensus clustering has emerged as an important elaboration of the classical clustering problem. Consensus clustering, also called aggregation of clustering (or partitions), refers to the situation in which a number of different (input) clusterings have been obtained for a particular dataset and it is desired to find a single (consensus) clustering which is a better fit in some sense than the existing clusterings. Consensus clustering is thus the problem of reconciling clustering information about the same data set coming from different sources or from different runs of the same algorithm. When cast as an optimization problem, consensus clustering is known as median partition, and has been shown to be NP-complete. Consensus clustering for unsupervised learning is analogous to ensemble learning in supervised learning.

Listed as below are some reason following for using consensus clustering [1].

- There are potential shortcomings for each of the known clustering techniques.
- Interpretations of results are difficult in a few cases.
- When there is no knowledge about the number of clusters, it becomes difficult.
- They are extremely sensitive to the initial settings.
- Some algorithms can never undo what was done previously.
- Iterative descent clustering methods, such as the SOM and K-Means clustering circumvent some of the shortcomings of hierarchical clustering by providing for univocally defined clusters and cluster boundaries. However, they lack the intuitive and visual appeal of hierarchical clustering, and the number of clusters must be chosen *a priori*.
- An extremely important issue in cluster analysis is the validation of the clustering results, that is, how to gain confidence about the significance of the clusters provided by the clustering technique (cluster numbers and cluster assignments). Lacking an external objective criterion (the equivalent of a known class label in supervised learning) this validation becomes somewhat elusive.

The advantages of consensus clustering are listed as below:

- Provides for a method to represent the consensus across multiple runs of a clustering algorithm, to determine the number of clusters in the data, and to assess the stability of the discovered clusters.
- The method can also be used to represent the consensus over multiple runs of a clustering algorithm with random restart (such as K-means, model-based Bayesian clustering, SOM, etc.), so as to account for its sensitivity to the initial conditions.
- It also provides for a visualization tool to inspect cluster number, membership, and boundaries.
- It is possible to extract lot of features/attributes from multiple runs of different clustering algorithms on the data. These features can give us valuable information in doing a final consensus clustering.

4.6.1 Consensus Clustering Framework

We are given a set of M data objects $D = \{x_1, x_2, ..., x_M\}$ and a set of P clusterings $\Pi = \{\pi_1, \pi_2, ..., \pi_P\}$ of the data objects in D. Each clustering π_p, $p = 1, ..., P$ is a mapping from D to $\{1, ..., n_{\pi p}\}$ where $n_{\pi p}$ is the number of clusters in π_p. The problem of clustering consensus is to find a new clustering π^* of the data set D that best summarizes the clustering ensemble Π.

Algorithm 4.13 shows the framework of consensus clustering. The consensus function is the main step in any clustering ensemble algorithm. Precisely, the great challenge in clustering ensemble is the definition of an appropriate consensus function, capable of improving the results of single clustering algorithms. In this step, the final data partition or consensus partition π^*, which is the result of any clustering ensemble algorithm, is obtained. However, the consensus among a set of clusterings is not obtained in the same way in all cases. There are two main consensus function approaches: objects co-occurrence and median partition. In the first approach, the idea is to determine which must be the cluster label

---

Algorithm 4.13: the framework of consensus clustering

Input: Data set D, clustering set Π, desired number of clusters K

Output: the consensus clustering results.

---

(1) define a consensus function
(2) optimize the consensus function until convergence
(3) return the consensus clustering result

---

associated to each object in the consensus partition. To do that, it is analyzed how many times an object belongs to one cluster or how many times two objects belong together to the same cluster. The consensus is obtained through a voting process among the objects. Somehow, each object should vote for the cluster to which it will belong in the consensus partition. This is the case, for example, of Relabeling and Voting and Co-association Matrix based methods. In the second consensus function approach, the consensus partition is obtained by the solution of an optimization problem, the problem of finding the median partition with respect to the cluster ensemble. Formally, the median partition is defined as:

$$\pi^* = \arg\max_{\pi \in \Pi} \sum_{p=1}^{P} \Gamma(\pi, \pi_p), \pi \neq \pi_p$$

where $\Gamma()$ is a similarity measure between partitions. The median partition is defined as the partition that maximizes the similarity with all partitions in the cluster ensemble. For example, Non-Negative Matrix Factorization and Kernel based methods follow this approach.

4.6.2 Some Consensus Clustering Methods

4.6.2.1 Relabeling and Voting-based Methods

The Relabeling and Voting methods are based on solving as first step the labeling correspondence problem and after that, in a voting process, the consensus partition is obtained. The labeling correspondence problem consists of the following: the label associated to each object in a partition is symbolic; there is no relation between the set of labels given by a clustering algorithm and the set of labels given by another one. The label correspondence is one of the main issues that make unsupervised combination difficult. The different clustering ensemble methods based on relabeling try to solve this problem using different heuristics such as bipartite matching and cumulative voting. Lots of method have been proposed: A general formulation for the voting problem as a multi-response regression problem was presented by Ayad and Kamel [13]. Plurality Voting (PV) [25], Voting-Merging (V-M) [59], Voting for fuzzy clusterings [21], Voting Active Clusters (VAC) [55], Cumulative Voting (CV) [12] and the methods proposed by Zhou and Tang [43] and Gordon and Vichi [27]. If a relation exists among the labels associated for each clustering algorithm, the voting definition of the clustering ensemble problem would be the most appropriate. However, the labeling correspondence problem is what makes the combination of clusterings difficult. This correspondence problem can only be solved, with certain accuracy, if all partitions have the same number of clusters. We consider this to be a strong restriction to the cluster ensemble problem. Then, in general, they are not recommended when the number of clusters in all partitions in the ensemble is not the same. Besides, very frequently, they could have high computational cost since the Hungarian algorithm to solve the label correspondence problem is $O(k^3)$, where k is the number of clusters in the consensus partition. On the other hand, these kinds of algorithms are usually easy to understand and implement.

4.6.2.2 Graph and Hyper Graph based Methods

This kind of clustering ensemble methods transform the combination problem into a graph or hyper graph partitioning problem. The difference among these methods lies on the way the (hyper)graph is built from the set of clusterings and how the cuts on the graph are defined in order to obtain the consensus partition. Strehl and Ghosh defined the consensus partition as the partition that most information shares with all partitions in the cluster ensemble [54]. Cluster-based Similarity Partitioning Algorithm (CSPA) [2], Hyper Graphs Partitioning Algorithm (HGPA) [3], Meta-CLustering Algorithm (MCLA) [4], Hybrid Bipartite Graph Formulation (HBGF) [24] are also the efficient graph and hyper graph-based consensus clustering method in the references. We consider that the main weakness of these kind of clustering ensemble methods is that they are not rigourously well-founded as a solution for the consensus clustering problem, in the sense that most of them are proposed as a solution for the median partition problem defined with the NMI similarity measure, but in practice, they are not solving this problem. These methods are more related with the object co-occurrence approach since in the (hyper)graph construction and in the partitioning algorithm, the relationship between individual objects are implicitly taken into account. In addition to that, these methods need a (hyper)graph partitioning algorithm in the final step, therefore, if we change this algorithm, the final result could change. Regardless of the fact that METIS and HMETIS are the most used algorithm for the (hyper)graph partitioning, they are not the only graph partitioning algorithm and they do not have to achieve the best results in all situations.

4.7 Chapter Summary

Clustering is an important data mining tools and it has been used in lots of application areas, such as, Biology, Information Retrieve, Climate, Psychology, Medicine and Business. In this chapter, we classify the proposed clustering algorithms into five categories: traditional clustering algorithm, high dimensional clustering algorithm and constraint-based clustering algorithm and consensus clustering algorithm. The traditional data clustering approaches include partitioning methods, hierarchical methods, density-based methods, grid-based methods, and model-based methods. Two different kinds of high-dimensional clustering algorithms have been described. In the constraint-based clustering algorithm subsection, we first discussed the concept of constraint-based clustering algorithm and then three traditional constraint-based clustering algorithms were introduced. The consensus clustering algorithm is based on the clustering results and it is a new way to find robust clustering result. In this chapter, we introduced the main frame work of consensus clustering at first, and then, we discussed the consensus function in detail. Eventually, two kinds of consensus clustering algorithms were introduced.

References

[1] http://en.wikipedia.org/wiki/Consensus clustering.

[2] http://www.lans.ece.utexas.edu/ strehl/diss/node80.html.

[3] http://www.lans.ece.utexas.edu/ strehl/diss/node81.html.

[4] http://www.lans.ece.utexas.edu/ strehl/diss/node82.html.

[5] E. Achtert, C. Bohm and P. Kroger. *DeLi-Clu: Boosting Robustness, Completeness, Usability, and Efficiency of Hierarchical Clustering by a Closest Pair Ranking.* 2006.

[6] E. Achtert, C. Bohm, H. peter Kriegel, P. Kroger and A. Zimek. On Exploring Complex Relationships of Correlation Clusters. *In: Statistical and Scientific Database Management*, pp. 7–7, 2007.

[7] C. C. Aggarwal, C. M. Procopiuc, J. L. Wolf, P. S. Yu and J. S. Park. Fast Algorithms for Projected Clustering. *Sigmod Record*, 28: 61–72, 1999.

[8] C. C. Aggarwal and P. S. Yu. Finding generalized projected clusters in high dimensional spaces. *Sigmod Record*, pp. 70–81, 2000.

[9] R. Agrawal, J. E. Gehrke, D. Gunopulos and P. Raghavan. Automatic subspace clustering of high dimensional data for data mining applications. *Sigmod Record*, 27: 94–105, 1998.

[10] R. Agrawal and R. Srikant. Fast algorithms for mining association rules. *In: Proc. of 20th Intl. Conf. on VLDB*, pp. 487–499, 1994.

[11] M. Ankerst, M. M. Breunig, H. peter Kriegel and J. Sander. *Optics: Ordering points to identify the clustering structure.* pp. 49–60. ACM Press, 1999.

[12] H. G. Ayad and M. S. Kamel. Cumulative Voting Consensus Method for Partitions with Variable Number of Clusters. *IEEE Transactions on Pattern Analysis and Machine Intelligence*, 30: 160–173, 2008.

[13] H. G. Ayad and M. S. Kamel. On voting-based consensus of cluster ensembles. Pattern Recognition, 43: 1943–1953, 2010.

[14] C. Baumgartner, C. Plant, K. Kailing, H. -P. Kriegel and P. Kroger. Subspace selection for clustering high-dimensional data. *In: Proc. 4th IEEE Int. Conf. on Data Mining (ICDM04)*, pp. 11–18, 2004.

[15] P. Berkhin. Survey of clustering data mining techniques. Technical report, 2002.

[16] C. Bohm, K. Kailing, H. Peter Kriegel and P. Kroger. Density Connected Clustering with Local Subspace Preferences. *In: IEEE International Conference on Data Mining*, pp. 27–34, 2004.

[17] M. Bilenko, S. Basu and R. J. Mooney. Integrating constraints and metric learning in semi-supervised clustering. *In: ICML*, pp. 81–88, 2004.

[18] Z. I. Botev, J. F. Grotowski and D. P. Kroese. Kernel density estimation via diffusion. *Annals of Statistics*, 38: 2916–2957, 2010.

[19] G. Celeux and G. Govaert. A classification EM algorithm for clustering and two stochastic versions. *Computational Statistics & Data Analysis*, 14: 315–332, 1992.

[20] A. P. Dempster, N. M. Laird and D. B. Rubin. Maximum likelihood from incomplete data via the em algorithm. *Journal of the Royal Statistical Society, Series B*, 39(1): 1–38, 1977.

[21] E. Dimitriadou, A. Weingessel and K. Hornik. A Combination Scheme for Fuzzy Clustering. *International Journal of Pattern Recognition and Artificial Intelligence*, 16: 332–338, 2002.

[22] M. Ester, H. Peter Kriegel, J. Sander and X. Xu. A Density-based Algorithm for Discovering Clusters in Large Spatial Databases with Noise. *In: Knowledge Discovery and Data Mining*, pp. 226–231, 1996.

[23] U. Fayyad, G. Piatetsky-shapiro and P. Smyth. From data mining to knowledge discovery in databases. *AI Magazine*, 17: 37–54, 1996.

[24] X. Z. Fern and C. E. Brodley. Solving cluster ensemble problems by bipartite graph partitioning. *In: Proceedings of the International Conference on Machine Learning*, 2004.

[25] B. Fischer and J. M. Buhmann. Bagging for Path-Based Clustering. *IEEE Transactions on Pattern Analysis and Machine Intelligence*, 25: 1411–1415, 2003.

[26] C. Fraley and A. E. Raftery. Model-based clustering, discriminant analysis, and density estimation. *Journal of the American Statistical Association*, 97: 611– 631, 2000.

[27] A. D. Gordon and M. Vichi. Fuzzy partition models for fitting a set of partitions. *Psy-chometrika*, 66: 229–247, 2001.

[28] N. Grira, M. Crucianu and N. Boujemaa. Active semi-supervised fuzzy clustering for image database categorization. *In:* ACM *Multimedia Conference*, pp. 9–16, 2005.

[29] P. J. Han J., Kamber M. Data Mining: Concepts and Techniques (3rd edition), 2012.

[30] y. . . m. . . Han J. and Kamber M., title = Data Mining: Concepts and Techniques.

[31] P. K. Hans-peter Kriegel and A. Zimek. Clustering high-dimensional data: A survey on subspace clustering, pattern-based clustering, and correlation clustering. *ACM Transactions on Knowledge Discovery From Data*, 3: 1–58, 2009.

[32] A. Hinneburg and D. A. Keim. An Efficient Approach to Clustering in Large Multimedia Databases with Noise. *In: Knowledge Discovery and Data Mining*, pp. 58–65, 1998.

[33] Z. hua Zhou and W. Tang. Clusterer ensemble. *Knowledge Based Systems*, 19: 77–83, 2006.

[34] C. hung Cheng, A. Fu, Y. Zhang, A. Wai-chee and F. Y. Zhang. Entropy-based subspace clustering for mining numerical data. pp. 84–93, 1999.

[35] T. P. P. K. H.-P. K. I. Ntoutsi and A. Zimek. Density-based projected clustering over high dimensional data streams. In *Proceedings of the 12th SIAM International Conference on Data Mining (SDM)*, pp. 21–28, 2012.

[36] M. M. B. Ismail and H. Frigui. Image annotation based on constrained clustering and seminaive bayesian model. *In: International Symposium on Computers and Communications*, pp. 431–436, 2009.

[37] A. K. Jain and R. C. Dubes. *Algorithms for Clustering Data*, 1988.

[38] E. Ka, K. Ng, A. W. chee Fu and R. C. wing Wong. Projective clustering by histograms. *IEEE Transactions on Knowledge and Data Engineering*, 17: 369–383, 2005.

[39] K. Kailing, H. -P. Kriegel, P. Kroeger and S. Wanka. Ranking interesting subspaces for clustering high-dimensional data. *In: PKDD*, pp. 241–252, 2003.

[40] L. Kaufman and P. J. Rousseeuw. *Finding Groups in Data: An Introduction to Cluster Analysis*, 1990.

[41] J. Liu and W. Wang. OP-Cluster: Clustering by Tendency in High-Dimensional Space. *In: IEEE International Conference on Data Mining*, pp. 187–194, 2003.

[42] H. -P. K. Michael, E. Houle, P. Kroger, E. Schubert and A. Zimek. *Can Shared-Neighbor Distances Defeat the Curse of Dimensionality?* 2010.

[43] G. Moise and J. Sander. 2008. Finding non-redundant, statistically significant regions in high-dimensional data: a novel approach to projected and subspace clustering. *In: Knowledge Discovery and Data Mining*, pp. 533–41, 2008.

[44] G. Moise, J. Sander and M. Ester. P3C: A Robust Projected Clustering Algorithm. *In: IEEE International Conference on Data Mining*, pp. 414–25, 2006.

[45] H. Nagesh, S. Goil and A. Choudhary. Adaptive grids for clustering massive data sets. *In: Proceedings of the 1st SIAM International Conference on Data Mining*, 2001.

[46] R. T. Ng and J. Han. CLARANS: A Method for Clustering Objects for Spatial Data Mining. *IEEE Transactions on Knowledge and Data Engineering*, 14: 1003–16, 2002.

[47] H. Peter Kriegel, P. Kroger, M. Renz and S. Wurst. A generic framework for efficient subspace clustering of high-dimensional data. *In: PROC. ICDM*, pp. 250–57. IEEE Computer Society, 2005.

[48] A. M. F. Petros Drineas, R. Kannan, S. Vempala and V. Vinay. Clustering Large Graphs via the Singular Value Decomposition. *Machine Learning*, 56: 9–33, 2004.

[49] C. M. Procopiuc, M. Jonesy, P. K. Agarwal and T. M. Muraliy. A Monte Carlo algorithm for fast projective clustering. pp. 418–427. ACM Press, 2002.

[50] E. Schikuta. Grid-clustering: an efficient hierarchical clustering method for very large data sets. *In: International Conference on Pattern Recognition*, Vol. 2, 1996.

[51] K. Sequeira and M. Zaki. SCHISM: a new approach to interesting subspace mining. *In: International Journal of Business Intelligence and Data Mining*, 1: 137–160, 2005.

[52] G. Sheikholeslami, S. Chatterjee and A. Zhang. WaveCluster: A Wavelet Based Clustering Approach for Spatial Data in Very Large Databases. *The Vldb Journal*, 8: 289–304, 2000.

[53] K. Sim, V. Gopalkrishnan, A. Zimek and G. Cong. A survey on enhanced subspace clustering. *Data Mining and Knowledge Discovery*, pp. 1–66, 2012.

[54] A. Strehl and J. Ghosh. Cluster Ensembles—A Knowledge Reuse Framework for Combining Multiple Partitions. *Journal of Machine Learning Research*, 3: 583–617, 2002.

[55] K. Tumer and A. K. Agogino. Ensemble clustering with voting active clusters. *Pattern Recognition Letters*, 29: 1947–1953, 2008.

[56] R. Vidal. Subspace Clustering. *IEEE Signal Processing Magazine*, 28: 52–68, 2011.

[57] W. Wang, J. Yang and R. Muntz. Sting: A statistical information grid approach to spatial data mining, pp. 186–95. Morgan Kaufmann, 1997.

[58] C. S. Warnekar and G. Krishna. A heuristic clustering algorithm using union of overlapping pattern-cells. *Pattern Recognition*, 11: 85–93, 1979.

[59] A. Weingessel, E. Dimitriadou and K. Hornik. Voting-merging: An ensemble method for clustering. icann. *In: Proc. Int. Conf. on Artificial Neural Networks*, pp. 217–224. Springer Verlag, 2001.

[60] M. L. Yiu and N. Mamoulis. Iterative projected clustering by subspace mining. *IEEE Transactions on Knowledge and Data Engineering*, 17: 176–189, 2005.

CHAPTER 5

Classification

Two common data mining techniques for finding hidden patterns in data are clustering and classification analysis. Although classification and clustering are often mentioned in the same breath, they are different analytical approaches. Imaging a database of customer records, where each record represents a customer's attributes. These can include identifiers such as name and address, demographic information such as gender and age, and financial attributes such as income and revenue spent. Clustering is an automated process to group related records together. Related records are grouped together on the basis of having similar values for attributes. This approach of segmenting the database via clustering analysis is often used as an exploratory technique because it is not necessary for the analyst to specify ahead of time how records should be related together. In fact, the objective of the analysis is often to discover clusters, and then examine the attributes and values that define the clusters or segments. As such, interesting and surprising ways of grouping customers together can become apparent, and this in turn can be used to drive marketing and promotion strategies to target specific types of customers. Classification is a different technique from clustering. It is similar to clustering in that it also segments customer records into distinct segments called classes. But unlike clustering, a classification analysis requires that the analyst know ahead of time how classes are defined. For example, classes can be defined to represent the likelihood that a customer defaults on a loan (Yes/No). It is necessary that each record in the dataset used to build the classifier already have a value for the attribute used to define classes. Because each record has a value for the attribute used to define the classes, and because the end-user decides on the attribute to use, classification is much less exploratory than clustering.

The objective of a classifier is not to explore the data to discover interesting segments, but rather to decide how new records should be classified—i.e., is this new customer likely to default on the loan? Classification routines in data mining also use a variety of algorithms

—and the particular algorithm used can affect the way records are classified. A common approach for classifiers is to use decision trees to partition and segment records. New records can be classified by traversing the tree from the root through branches and nodes, to a leaf representing a class. The path a record takes through a decision tree can then be represented as a rule. For example, Income<$30,000 and age<25, and debt=High, then Default Class=Yes. But due to the sequential nature of the way a decision tree splits records (i.e., the most discriminative attribute-values [e.g., Income] appear early in the tree) can result in a decision tree being overly sensitive to initial splits. Therefore, in evaluating the goodness of fit of a tree, it is important to examine the error rate for each leaf node (proportion of records incorrectly classified). A nice property of decision tree classifiers is that because paths can be expressed as rules, then it becomes possible to use measures for evaluating the usefulness of rules such as Support, Confidence and Lift to also evaluate the usefulness of the tree. Although clustering and classification are often used for purposes of segmenting data records, they have different objectives and achieve their segmentations through different ways. Knowing which approach to use is important for decision-making.

5.1 Classification Definition and Related Issues

The data analysis task classification is where a model or classifier is constructed to predict categorical labels (the class label attribute). For example, Categorical labels include "safe" or "risky" for the loan application data. In general, data classification includes the following two-step process. Step 1: A classifier is built describing a predetermined set of data classes or concepts. This is the learning step (or training phase), where a classification algorithm builds the classifier by analyzing or "learning from" a training set made up of database tuples and their associated class labels. Each tuple, is assumed to belong to a predefined class called the class label attribute. Because the class label of each training tuple is provided, this step is also known as supervised learning. The first step can also be viewed as the learning of a mapping or function, $y = f(X)$, that can predict the associated class label y of a given tuple X. Typically, this mapping is represented in the form of classification rules, decision trees, or mathematical formulae. In step 2, the model is used for classification.

The predictive accuracy of the classifier is very important and should be estimated at first. If we were to use the training set to measure the accuracy of the classifier, this estimate would likely be optimistic, because the classifier tends to overfit the data. Therefore, a test set is used, made up of test tuples and their associated class labels. The associated class label of each test tuple is compared with the learned classifier's class prediction for that tuple. If the accuracy of the classifier is considered acceptable, the classifier can be used to classify future data tuples for which the class label is not known. For example, the classification rules learned in Fig. from the analysis of data from previous loan applications can be used to approve or reject new or future loan applicants. The preparing of the data and the quality of a classifier are two important regarding issues of classification. The following preprocessing steps may be applied to the data to help improve the accuracy, efficiency, and scalability of the classification process.

Data cleaning: This refers to the preprocessing of data in order to remove or reduce noise and the treatment of missing values. This step can help reduce confusion during learning.

Relevance analysis: Many of the attributes in the data may be redundant. A database may also contain irrelevant attributes. Hence, relevance analysis in the form of correlation analysis and attribute subset selection, can be used to detect attributes that do not contribute to the classification or prediction task.

Data transformation and reduction: Normalization involves scaling all values for a given attribute so that they fall within a small specified range, such as 0:0 to 1:0.

The data can also be transformed by generalizing it to higher-level concepts. Concept hierarchies may be used for this purpose. Data can also be reduced by applying many other methods, ranging from wavelet transformation and principle components analysis to discretization techniques, such as binning, histogram analysis, and clustering. Ideally, the time spent on relevance analysis, when added to the time spent on learning from the resulting "reduced" attribute subset, should be less than the time that would have been spent on learning from the original set of attributes. Hence, such analysis can help improve classification efficiency and scalability.

Classification methods can be compared and evaluated according to the following criteria:

- *Accuracy*: The accuracy of a classifier refers to the ability of a given classifier to correctly predict the class label of new or previously unseen data. Estimation techniques are cross-validation and bootstrapping. Because the accuracy computed is only an estimate of how well the classifier or predictor will do on new data tuples, confidence limits can be computed to help gauge this estimate.
- *Speed*: This refers to the computational costs involved in generating and using the given classifier.

- *Robustness*: This is the ability of the classifier to make correct predictions given noisy data or data with missing values.
- *Scalability*: This refers to the ability to construct the classifier efficiently given large amounts of data.
- *Interpretability*: This refers to the level of understanding and insight that is provided by the classifier.
- *End nodes*: represented by triangles.

5.2 Decision Tree and Classification

This section introduces decision tree first, and then discusses a decision tree classifier.

5.2.1 Decision Tree

A decision tree is a decision support tool that uses a tree-like graph or model of decisions and their possible consequences, including chance event outcomes, resource costs, and utility. It is one way to display an algorithm. Decision trees are commonly used in operations research, specifically in decision analysis, to help identify a strategy most likely to reach a goal. If in practice decisions have to be taken online with no recall under incomplete knowledge, a decision tree should be paralleled by a probability model as a best choice model or online selection model algorithm. Another use of decision trees is as a descriptive means for calculating conditional probabilities. In general, a "decision tree" is used as a visual and analytical decision support tool, where the expected values (or expected utility) of competing alternatives are calculated. A decision tree consists of three types of nodes:

- Decision nodes—commonly represented by squares.
- Chance nodes—represented by circles.
- End nodes—represented by triangles.

Commonly, a decision tree is drawn using flow chart symbols as it is easier for many to read and understand. Figure 5.2.1 shows a decision tree which is drawn using flow chart symbols. A decision tree has only burst nodes (splitting paths) but no sink nodes (converging paths). Therefore, used manually, they can grow very big and are then often hard to draw fully by hand. Traditionally, decision trees have been created manually—as the aside example shows—although increasingly, specialized software is employed. Decision trees have several advantages:

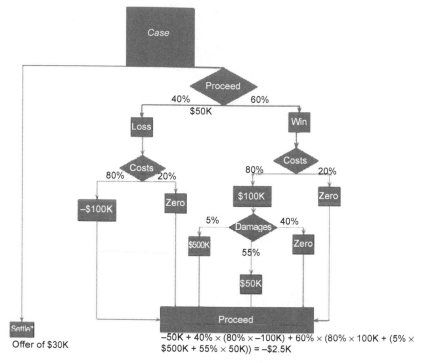

Figure 5.2.1: An example of a decision tree.

- Are simple to understand and interpret. People are able to understand decision tree models after a brief explanation.
- Have value even with little hard data. Important insights can be generated based on experts describing a situation (its alternatives, probabilities, and costs) and their preferences for outcomes.
- Possible scenarios can be added.
- Worst, best and expected values can be determined for different scenarios. Use a white box model. If a given result is provided by a model.
- Can be combined with other decision techniques.

Like other methods, decision tree also has some disadvantages. These include:

- For data including categorical variables with different number of levels, information gain in decision trees are biased in favor of those attributes with more levels [10].
- Calculations can get very complex particularly if many values are uncertain and/or if many outcomes are linked.

5.2.2 Decision Tree Classification

Decision tree classification uses a decision tree as a predictive model which maps observations about an item to conclusions about the item's target value. More descriptive names for such tree models are classification trees or regression trees. In these tree structures, leaves represent class labels and branches represent conjunctions of features that lead to those class labels. In decision analysis, a decision tree can be used to visually and explicitly represent decisions and decision making. In data mining, a decision tree describes data but not decisions; rather the resulting classification tree can be an input for decision making. This page deals with decision trees in data mining. Decision tree learning is a method commonly used in data mining. The goal is to create a model that predicts the value of a target variable based on several input variables. An example is shown on the right. Each interior node corresponds to one of the input variables; there are edges to children for each of the possible values of that input variable. Each leaf represents a value of the target variable given the values of the input variables represented by the path from the root to the leaf. A tree can be "learned" by splitting the source set into subsets based on an attribute value test. This process is repeated on each derived subset in a recursive manner called recursive partitioning. The recursion is completed when the subset at a node has all the same value of the target variable, or when splitting no longer adds value to the predictions. This process of top-down induction of decision trees (TDIDT) [14] is an example of a greedy algorithm, and it is by far the most common strategy for learning decision trees from data, but it is not the only strategy. In fact, some approaches have been developed recently allowing tree induction to be performed in a bottom-up fashion [4].

In classification, there have three different nodes in decision tree which are described as following:

- A root node that has no incoming edges and zero or more outgoing edges.
- Internal nodes, each of which has exactly one incoming edge and two or more outgoing edges.
- Leaf or terminal nodes, each of which has exactly one incoming edge and no outgoing edges.

Figure 5.2.2 shows the decision tree for the survival passengers on the Titanic classification problem. In the figure, "sibsp" is the number of spouses or siblings aboard. Each interior node corresponds to one of the input variables; there are edges to children for each of the possible values of that input variable. Each leaf represents a value of the target variable given the values of the input variables represented by the path from the root to the leaf.

5.2.3 Hunt's Algorithm

To build an optimal decision tree is the key problem in a decision tree classifier. In general, decision trees can be constructed from a given set of attributes. While some of the trees are more accurate than others, finding the optimal tree is computationally infeasible because of the exponential size of the search space. However, various efficient algorithms have been developed to construct a reasonably accurate, albeit suboptimal, decision tree in a reasonable amount of time. These algorithms usually employ a greedy strategy that grows a decision tree by making a series of locally optimum decisions about which attribute to use for partitioning the data. Hunt's algorithm is one of the efficient method for constructing a decision tree. It grows a decision tree in a recursive fashion by partitioning the training records into successively purer subsets. Let D_t be the set of training records that reach a node t. The general recursive procedure is defined as algorithm 5.1 [17]. It recursively applies the procedure to each subset until all the records in the subset belong to the same class. Hunt's algorithm assumes that each combination of attribute sets has a unique class label during the procedure. If all the records associated with D_t have identical attribute values except for the class label, then it is not possible to split these records any further. In that case, the node is declared a leaf node with the same class label as the majority class of training records associated with this node.

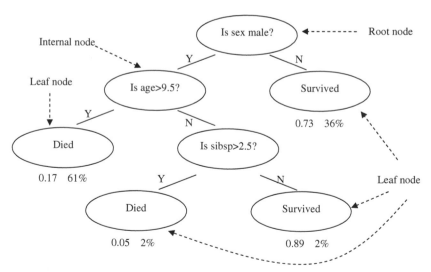

Figure 5.2.2: An example of decision tree classification for Titanic

---

Algorithm 5.1: Hunt's algorithm

---

(1) If D_t contains records that belong the same class y_t, then t is a leaf node labeled as y_t
(2) If D_t is an empty set, then t is a leaf node labeled by the default class, y_d
(3) If D_t contains records that belong to more than one class, use an attribute test to split the data into smaller subsets.

5.3 Bayesian Network and Classification

5.3.1 Bayesian Network

Bayesian network theory can be thought of as a fusion of incidence diagrams and Bayes' theorem. A Bayesian network, or belief network, shows conditional probability and causality relationships between variables. For example, a Bayesian network could represent the probabilistic relationships between diseases and symptoms. Given the symptoms, the network can be used to compute the probabilities of the presence of various diseases. The probability of an event occurring given that another event has already occurred is called *conditional probability*. The probabilistic model is described qualitatively by a directed acyclic graph, or DAG. The vertices of the graph, which represent variables, are called nodes. The nodes are represented as circles containing the variable name. The connections between the nodes are called arcs, or edges. The edges are drawn as arrows between the nodes, and represent dependence between the variables. Therefore, any pair of nodes indicates that one node is the parent of the other so there are no independent assumptions. Independent assumptions are implied in Bayesian networks by the absence of a link. Figure 5.3.1 shows an example of DAG. The node where the arc originates is called the *parent*, while the node where the arc ends is called the *child*. In this case, V_0 is a parent of V_1 and V_2, V_2 has parents V_0 and V_1. Nodes that can be reached from other nodes are called *descendants*. Nodes that lead a path to a specific node are called *ancestors*. For example, V_1 and V_2 are descendants of V_0, and V_1 is ancestors of V_2 and V_3. Since no child can be its own ancestor or descendent, there are no loops in Bayesian networks. Bayesian networks will generally also include a set of probability tables, stating the probabilities for the true/false values of the variables. The main point of Bayesian Networks is to allow for probabilistic inference to be performed. This means that the probability of each value of a node in the Bayesian network can be computed when the values of the other variables are known. Also, because independence among the variables is easy to recognize since conditional relationships are clearly defined by a graph edge, not all joint probabilities in the Bayesian system need to be calculated in order to make a decision. Classification with Bayesian network.

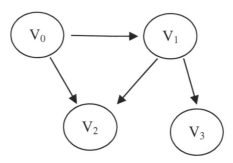

Figure 5.3.1: An example of DAG

Figure 5.3.2 depicts the possible structure of a Bayesian network used for classification. The dotted lines denote potential links, and the blue box indicates that additional nodes and links can be added to the model, usually between the input and output nodes.

In order to perform classification with a Bayesian network such as the one depicted in Fig. 5.3.2, first evidence must be set on the input nodes, and then the output nodes can be queried using standard Bayesian network inference. The result will be a distribution for each output node, so that you can not only determine the most probable state for each output, but also see the probability assigned to each output state. Figure 5.3.3 shows the structure of a Naive Bayes classifier, which is the simplest form of useful Bayesian network classifier. The links in a Naive Bayes model are directed from output to input, which gives the model its simplicity, as there are no interactions between the inputs, except indirectly via the output. Note however that directing links from output to input, is not a requirement for all Bayesian network classifiers.

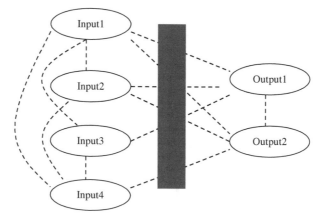

Figure 5.3.2: Generic structure of a Bayesian network classifier

One of the most effective classifiers, in the sense that its predictive performance is competitive with state-of-the-art classifiers, is the so-called naive Bayesian classifier described, for example, by Duda and Hart [9] and by Langley et al. [12]. This classifier learns from training data the conditional probability of each attribute Ai given the class label C. Classification is then done by applying Bayes rule to compute the probability of C given the particular instance of $A_1,...,A_n$, and then predicting the class with the highest posterior probability. This computation is rendered feasible by making a strong independence assumption: all the attributes Ai are conditionally independent given the value of the class C.

5.3.2 Backpropagation and Classification

5.3.2.1 Backpropagation Method

Backpropagation [1] is a common method of training artificial neural networks so as to minimize the objective function. Arthur E. Bryson and Yu-Chi Ho described it as a multi-stage dynamic system optimization method in 1969 [15, 6]. It wasn't until 1974 and later, when applied in the context of neural networks and through the work of Paul Werbos [19], Rumelhart and Kubat [16, 11], that it gained recognition, and it led to a "renaissance" in the field of artificial neural network research. It is a supervised learning method, and is a generalization of the delta rule. It requires a dataset of the desired output for many inputs, making up the training set. It is most useful for feed-forward networks (networks that have no feedback, or simply, that have no connections that loop). The term is an abbreviation for "backward propagation of errors". Backpropagation requires that the activation

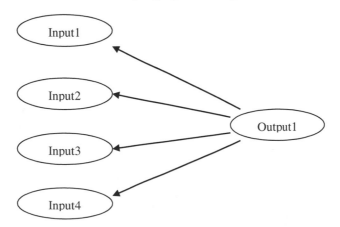

Figure 5.3.3: Naive Bayes model

function used by the artificial neurons (or "nodes") be differentiable. The main framework of Backpropagation could be described as Algorithm 5.2. In step 2.2, the ratio influences the speed and quality of learning; it is called the learning rate. The sign of the gradient of a weight indicates where the error is increasing; this is why the weight must be updated in the opposite direction.

Algorithm 5.2: Backpropagation

---

(1) Propagation

      (1.1) Forward propagation of a training pattern's input through the neural network in order to generate the propagation's output activations.

      (1.2) Backward propagation of the propagation's output activations through the neural network using the training pattern's target in order to generate the deltas of all output and hidden neurons.

(2) Weight update

      (2.1) Multiply its output delta and input activation to get the gradient of the weight.

      (2.2) Bring the weight in the opposite direction of the gradient by subtracting a ratio of it from the weight.

(3) Repeat phases 1 and 2 until the performance of the network is satisfactory.

5.3.2.2 Classifier with Backpropagation

The model structure of BP (backpropagation) classification algorithm uses full connection each layers and nodes from input layer to output layer. Obviously, it needs a lot of calculation. However, we are not still satisfied with standard neural network or back-propagation model based decision support system because we want to get better quality of decision performance and less computing iteration when we want to develop in a specific domain area.

5.3.3 Association-based Classification

Association rule mining is an important and highly active area of data mining research. Recently, data mining techniques have been developed that apply concepts used in association rule mining to the problem of classification. In this section, we study three methods in historical order. The first two, ARCS_ORCS [2] and associative classification [18], use association rules for classification. The third method, CAEP [8], mines "emerging patterns" that consider the concept of support used in mining associations. The first method mines association rules based on clustering and then employs the rules for classification. The ARCS or Association Rule Clustering System, mines association rules of the form $Aquan_1 ... Aquan_2 = \xi$

Acat where Aquan₁ and Aquan₂ are tests on quantitative attributive ranges (where the ranges are dynamically determined), and Acat assigns a class label for a categorical attribute from the given training data. Association rules are plotted on a 2-D grid. The algorithm scans the grid, searching for rectangular clusters of rules. In this way, adjacent ranges of the quantitative attributes occurring within a rule cluster may be combined. The clustered association rules generated by ARCS were empirically found to be slightly more accurate than C4.5 when there are outliers in the data. The accuracy of ARCS is related to the degree of discretization used. In terms of scalability, ARCS requires "a constant amount of memory", regardless of the database size. C4.5 has exponentially higher execution times than ARCS, requiring the entire database, multiplied by some factor, to fit entirely in main memory. The second method is referred to as associative classification. It mines rules of the form condset= $>y$, where condset is a set of items (or attribute-value pairs) and y is a class label. Rules that satisfy pre-specified minimum supports are frequent, where a rule has support s. if $s\%$ of the samples in the given data set contain consent and belong to class y. A rule satisfying minimum confidence is called accurate, where a rule has confidence c, if $c\%$ of the samples in the given data set that contain consent belong to class y. If a set of rules has the same consent, then the rule with the highest confidence is selected as the possible rule (PR) to represent the set.

The association classification method consists of two steps. The first step finds the set of all PRs that are both frequent and accurate. It uses an iterative approach, where prior knowledge is used to prune the rule search. The second step uses a heuristic method to construct the classifier, where the discovered rules are organized according to decreasing precedence based on their confidence and support. The algorithm may require several passes over the data set, depending on the length of the longest rule found. When classifying a new sample, the first rule satisfying the sample is used to classify it. The classifier also contains a default rule, having lowest precedence, which specifies a default class for any new sample that is not satisfied by any other rule in the classifier. In general, the associative classification method was empirically found to be more accurate than C4.5 on several data sets. Each of the above two steps was shown to have linear scale-up.

The third method, CAEP (classification by aggregating emerging patterns), uses the notion of itemset supports to mine emerging patterns (EPs), which are used to construct a classifier. Roughly speaking, an EP is an itemset (or set of items) whose support increases significantly from one class of data to another. The ratio of the two supports is called the growth rate of the EP. For example, suppose that we have a data set of customers with the classes buyscomputer = "yes", or C1, and buys computer = "no", or C2, the itemset age = "≤30", student = "no" is a typical EP, whose support increases from 0.2% in C1 to 57.6% in C2 at a growth rate of EP = 288. Note that an item is either a simple equality test; on a categorical attribute is in an interval. Each EP is a multi-attribute test and can be very strong at differentiating instances of one class from another. For instance, if a new sample X contains the above EP, then with odds of 99.6% we can claim that X belongs to C2. In general, the differentiating power of an EP is roughly proportional to its growth rate and its support in the target class.

For each class C, CAEP find EPs satisfying given support and growth rate thresholds, where growth rate I computed with respect to the set of all non-C samples versus the target set of all C samples, "Borderbased" algorithms can be used for this purpose. Where classifying a new sample, X, for each class C, the differentiating power of the EPs of class C that occur in X are aggregated to derive a score for C that is then normalized. The class with the largest normalized score determines the class label of X. CAEP has been found to be more accurate than C4.5 and association0-based classification on several data sets. It also performs well on data sets where the mail class of interest is in the minority. It scales up on data volume and dimensionality. An alternative classifier, called the JEP-classifier, was proposed based on jumping emerging patterns (JEPs). A JEP is a special type of EP, defined as an itemset whose support increases abruptly from zero in one data set to nonzero in another data set. The two classifiers are considered complementary.

5.3.4 Support Vector Machines and Classification

5.3.4.1 Support Vector Machines

In machine learning, support vector machines [7] are supervised learning models with associated learning algorithms that analyze data and recognize patterns, used for classification and regression analysis. The basic SVM takes a set of input data and predicts, for each given input, which of two possible classes forms the output, making it a non-probabilistic binary linear classifier. Given a set of training examples, each marked as belonging to one of two categories, an SVM training algorithm builds a model that assigns new examples into one category or the other. An SVM model is a representation of the examples as points in space, mapped so that the examples of the separate categories are divided by a clear gap that is as wide as possible. New examples are then mapped into that same space and predicted to belong to a category based on which side of the gap they fall on. In addition to performing linear classification, SVMs can efficiently perform non-linear classification using what is called the kernel trick, implicitly mapping their inputs into high-dimensional feature spaces. More formally, a support vector machine constructs a hyperplane or set of hyperplanes in a high or infinite-dimensional space, which can be used for classification, regression, or other tasks. Intuitively, a good separation is achieved by the hyperplane that has the largest distance to the nearest training data point of any class (so-called functional margin), since in general the larger the margin the lower the generalization error of the classifier. Whereas the original problem may be stated in a finite dimensional space, it often happens that the sets to discriminate are not linearly separable in that space. For this reason, it was proposed that the original finite-dimensional space be mapped into a much higher-dimensional space, presumably making the separation easier in that space. To keep the computational load reasonable, the mappings used by SVM schemes are designed to ensure that dot products may be computed easily in terms of the variables in the original space, by defining them in terms of a kernel function $K(x, y)$ selected to suit the problem [20]. The hyperplanes in the higher-dimensional space are defined as the set of points whose inner product with a vector in that space is constant. The vectors defining the hyperplanes can be chosen to be linear combinations with parameters of images of feature vectors that occur in the data base. With this choice of a hyperplane, the points in the feature space that are mapped into the hyperplane are defined by the relation: $\sum_i a_i K(x_i, x)$ = *constant*. Note that if $K(x, y)$ becomes small as y grows further away from x, each element in the sum measures the degree of closeness of the test point x to the corresponding data base point x_i. In this way, the sum of kernels above can be used to measure the relative nearness of each test point to the data points originating in one or the other of the sets to be discriminated. Note the fact that the set of points x mapped into any hyperplane can be quite convoluted as a result, allowing much more complex discrimination between sets which are not convex at all in the original space.

5.3.4.2 Classifier with Support Vector Machines

The original optimal hyperplane algorithm proposed by Vapnik in 1963 was a linear classifier. However, in 1992, Bernhard E. Boser, Isabelle M. Guyon and Vladimir N. Vapniks suggested a way to create nonlinear classifiers by applying the kernel trick (originally proposed by Aizerman et al. [3]) to maximum-margin hyperplanes [5]. The resulting algorithm is formally similar, except that every dot product is replaced by a nonlinear kernel function. This allows the algorithm to fit the maximum-margin hyperplane in a transformed feature space. The transformation may be nonlinear and the transformed space high dimensional; thus though the classifier is a hyperplane in the high-dimensional feature space, it may be nonlinear in the original input space. If the kernel used is a Gaussian radial basis function, the corresponding feature space is a Hilbert space of infinite dimensions.

Maximum margin classifiers are well regularized, so the infinite dimensions do not spoil the results. Some common kernels include:

- Polynomial (homogeneous): $K(x_i, x_j) = (x_i \cdot x_j)^d$.
- Polynomial (inhomogeneous): $K(x_i, x_j) = (x_i \cdot x_j + 1)^d$.
- Gaussian radial basis function: $K(x_i, x_j) = \exp(-\gamma \cdot | \, |x_i - x_j| \, |^2)$, for $\gamma > 0$.
- Hyperbolic tangent: $k(x_i, x_j) = \tanh(Kx_i \cdot x_j + c)$, for $k > 0 \; c < 0$.

The classifier with SVM has the following properties. SVMs belong to a family of generalized linear classifiers and can be interpreted as an extension of the perception. They can also be considered a special case of Tikhonov regularization. A special property is that they simultaneously minimize the empirical classification error and maximize the geometric margin; hence they are also known as maximum margin classifiers. A comparison of the SVM to other classifiers has been made by Meyer, Leisch and Hornik [13].

Simple feature selection algorithms are ad hoc, but there are also more methodical approaches. From a theoretical perspective, it can be shown that optimal feature selection for supervised learning problems requires an exhaustive search of all possible subsets of features of the chosen cardinality. If large numbers of features are available, this is impractical. For practical supervised learning algorithms, the search is for a satisfactory set of features instead of an optimal set. Feature selection algorithms typically fall into two categories: feature ranking and subset selection. Feature ranking ranks the features by a metric and eliminates all features that do not achieve an adequate score. Subset selection searches the set of possible features for the optimal subset. In statistics, the most popular form of feature selection is stepwise regression. It is a greedy algorithm that adds the best feature (or deletes the worst feature) at each round. The main control issue is deciding when to stop the algorithm. In machine learning, this is typically done by cross-validation. In statistics, some criteria are optimized. This leads to the inherent problem of nesting. More robust methods have been explored, such as branch and bound and piecewise linear network. Subset selection evaluates a subset of features as a group for suitability. Subset selection algorithms can be broken into Wrappers, Filters and Embedded. Wrappers use a search algorithm to search through the space of possible features and evaluate each subset by running a model on the subset. Wrappers can be computationally expensive and have a risk of over fitting to the model. Filters are similar to Wrappers in the search approach, but instead of evaluating against a model, a simpler filter is evaluated. Embedded techniques are embedded in and specific to a model. Many popular search approaches use greedy hill climbing, which iteratively evaluates a candidate subset of features, then modifies the subset and evaluates if the new subset is an improvement over the old. Evaluation of the subsets requires a scoring metric that grades a subset of features. Exhaustive search is generally impractical, so at some implementor (or operator) defined stopping point, the subset of features with the highest score discovered up to that point is selected as the satisfactory feature subset. The stopping criterion varies by algorithm; possible criteria include: a subset score exceeds a threshold, a program's maximum allowed run time has been surpassed, etc. Alternative search-based techniques are based on targeted projection pursuit which finds low-dimensional projections of the data that score highly: the features that have the largest projections in the lower dimensional space are then selected. The classification problem can be restricted to consideration of the two-class problem without loss of generality. In this problem the goal is to separate the two classes by a function which is induced from available examples. The goal is to produce a classifier that will work well on unseen examples.

5.4 Chapter Summary

In this chapter, we've talk about some methods which are proposed in classification. Decision trees and Bayesian Network (BN) generally have different operational profiles, when one is very accurate the other is not and vice versa. On the contrary, decision trees and rule classifiers have a similar operational profile. The goal of classification result integration algorithms is to generate more certain, precise and accurate system results. Numerous methods have been suggested for the creation of an ensemble of classifiers. Although or perhaps because many methods of ensemble creation have been proposed, there is as yet no clear picture of which method is best. Classification methods are typically strong in modeling interactions. Several of the classification methods produce a set of interacting loci that best predict the phenotype. However, a straightforward application of classification methods to large numbers of markers has a potential risk picking up randomly associated markers.

References

[1] http://en.wikipedia.org/wiki/Backpropagation.
[2] http://www.gov.bc.ca/citz/iao/records-mgmt/arcs-orcs/
[3] M. Aizerman, E. Braverman and L. Rozonoer. Theoretical foundations of the potential function method in pattern recognition learning, 1964.
[4] R. Barros. Fa bottom-up oblique decision tree induction algorithm. *In: 11th International Conference on Intelligent Systems Design and Applications (ISDA)*, 2011, pp. 450–456, 2011.
[5] B. E. Boser, I. M. Guyon and V. N. Vapnik. A training algorithm for optimal margin classifiers. *In: Computational Learning Theory*, pp. 144–152, 1992.

[6]   A. E. Bryson and Y. Ho. Applied optimal control: optimization, estimation, and control, 1975.

[7]   C. C. and V. V. Support-vector networks. *Machine Learning*, 20: 273–297, 2007.

[8]   G. Dong, X. Zhang and L. Wong. Caep: Classification by aggregating emerging patterns, pp 30–42, 1999.

[9]   R. O. Duda and P. E. Hart. *Pattern Classification and Scene Analysis*. John Wiley & Sons, New York, 1973.

[10] S. hyuk Cha. A genetic algorithm for constructing compact binary decision trees. Journal of *Pattern Recognition Research*, 4(1): 1–13, 2009.

[11] M. Kubat. *Introduction to Machine Learning*, 1992.

[12] P. Langley, W. Iba and K. Thompson. An analysis of bayesian classifiers. *In: Proceedings of the Tenth National Conference on Artificial Intel-Ligence*, pp. 223–28. MIT Press, 1992.

[13] A. H. K. Meyer D. and F. Leisch. The support vector machine under test. *Neurocomputing*, 55: 169–186, 2003.

[14] J. R. Quinlan. Induction of decision trees. *Mach. Learn.*, 1(1): 81–106, March 1986.

[15] S. R. and P. N. *Artificial Intelligence: A Modern Approach*. Prentice Hall, 2011.

[16] D. E. Rumelhart, G. E. Hinton and R. J. Williams. Learning Representations by Back-Propagating Error. *Nature*, 1988.

[17] P.-N. Tan, M. Steinbach and V. Kumar. *Introduction to Data Mining, (First Edition)*. Addison-Wesley Longman Publishing Co., Inc., Boston, MA, USA, 2005.

[18] F. A. Thabtah. A review of associative classification mining. *Knowledge Engineering Review*, 22: 37–65, 2007.

[19] P. Werbos. Beyond regression: new tools for prediction and analysis in the behavioral sciences, 1974.

[20] W. T. William H.P., Saul A.T. and B. P.F. *Numerical Recipes 3rd Edition: The Art of Scientific Computing*. Hardcover, 2007.

CHAPTER 6

Frequent Pattern Mining

Frequent pattern mining is one of the most fundamental research issues in data mining, which aims to mine useful information from huge volume of data. The purpose of searching such frequent patterns (i.e., association rules) is to explore the historical supermarket transaction data, which is indeed to discover the customer behavior based on the purchased items. Association rules present the fact that how frequently items are bought together. For example, an association rule "beer->diaper (75%)" indicates that 75% of the customers that bought beer also bought diaper. Such rules can be used to make prediction and recommendation for customers and store layout. Stemmed from the basic itemset data, rule discovery on more general and complex data (i.e., sequence, tree, graph) has been thoroughly explored in the past decade. In this chapter, we introduce the basic techniques of frequent pattern mining on different type of data, i.e., itemset, sequence, tree, and graph.

In the following sections, most classic algorithms and techniques for data mining will be introduced. Association rule mining will be presented in Section 6.1. Sequential pattern mining will be introduced in Section 6.2. Frequent tree and graph mining will be presented in Section 6.3 and Section 6.4, respectively. Chapter summary will be presented in Section 6.5.

6.1 Association Rule Mining

Data mining is to find valid, novel, potentially useful, and ultimately understandable patterns in data [18]. The most fundamental and important issue in data mining is association rule mining [1], which was first introduced in the early 1990s.

The purpose of searching association rules is to analyze the co-existence relation between items, which is then utilized to make appropriate recommendation. The issue has attracted a great deal of interest during the recent surge in data mining research because it is the basis of many applications, such as customer behavior analysis, stock trend prediction, and DNA sequence analysis. For example, an association rule "bread $\Rightarrow$ milk (90%)" indicates that nine out of ten customers who bought bread also bought milk. These rules can be useful for store layout, stock prediction, DNA structure analysis, and so forth.

Table 6.1: A database

Tid	Transaction
10	bread, milk
20	bread, chocolate, cookie
30	chocolate, cookie
40	milk
50	bread, cookie

6.1.1 Association Rule Mining Problem

The problem of association rule discovery can be stated as follows [1]: Let I = $\{i_1, i_2, \ldots, i_k\}$ be a set of items. A subset of I is called an *itemset*. The itemset, t_j, is denoted as $\{x_1, x_2 \ldots x_m\}$, where x_k is an item, i.e., $x_k \in I$ for $1 \leq k \leq m$. The number of items in an itemset is called the *length* of the itemset. An itemset with length l is called an l-itemset. An itemset, $t_a = \{a_1, a_2, \ldots, a_n\}$, is contained in another itemset, $t_b = \{b_1, b_2, \ldots, b_m\}$, if there exists integers 1 $\leq i_1 < i_2 < \ldots < i_n \leq m$, such that $a_1 \subseteq b_{i_1}, a_2 \subseteq b_{i_2}, \ldots, a_n \subseteq b_{i_n}$. We denote t_a a *subset* of t_b, and t_b a *superset* of t_a.

The *support* of an itemset X, denoted as $support(X)$, is the number of transactions in which it occurs as a subset. A k length subset of an itemset is called a k-subset. An itemset is *frequent* if its support is greater than a user-specified minimum support (min_{sup}) value. The set of frequent k-itemsets is denoted by F_k.

An association rule is an expression $A \Rightarrow B$, where A and B are itemsets. The support of the rule is given as $support(A \Rightarrow B) = support(A \cup B)$ and the *confidence* of the rule is given as $conf(A \Rightarrow B) = support(A \cup B)/support(A)$ (i.e., the conditional probability that a transaction contains B, given that it contains A). A rule is *confident* if its confidence is greater than a user-specified minimum confidence (min_{conf}).

The associate rule mining task is to generate all the rules, whose supports are greater than min_{sup}, and the confidences of the rules are greater than min_{conf}. The issue can be tackled by a two-stage strategy [2]:

- Find all frequent itemsets. This stage is the most time consuming part. Given k items, there can be potentially 2^k frequent itemsets. Therefore, almost all the works so far have focused on devising efficient algorithms to discover the frequent itemsets, while avoiding to traverse unnecessary search space somehow. In this chapter, we mainly introduce the basic algorithms on finding frequent itemsets.
- Generate confident rules. This stage is relatively straightforward and can be easily completed.

Almost all the association rule mining algorithms apply the two-stage rule discovery approach. We will discuss it in more detail in the next few sections.

Example 1. Let our example database be the database D shown in Table 6.1 with min_{sup}=1 and min_{conf}=30%. Table 6.2 shows all frequent itemsets in D. Table 6.3 illustrates all the association rules. For the sake of simplicity and without loss of generality, we assume that items in transactions and itemsets are kept sorted in the lexicographic order unless stated otherwise.

Table 6.2: Frequent itemsets

Frequent Itemset	Transactions	Support
bread	10, 20, 50	3
milk	10, 40	2
chocolate	20, 30	2
cookie	20, 30, 50	3
bread, milk	10	1
bread, chocolate	20	1
bread, cookie	20, 50	2
chocolate, cookie	20, 30	2
bread, chocolate, cookie	20	1

Table 6.3: Association rules

Association Rule	Support	Confidence
bread $\Rightarrow$ cookie	2	67%
milk $\Rightarrow$ bread	1	50%
chocolate $\Rightarrow$ bread	1	50%
chocolate $\Rightarrow$ cookie	2	100%
cookie $\Rightarrow$ bread	2	67%
cookie $\Rightarrow$ chocolate	2	67%
bread, chocolate $\Rightarrow$ cookie	1	100%
bread, cookie $\Rightarrow$ chocolate	1	50%
chocolate, cookie $\Rightarrow$ bread	1	50%
chocolate $\Rightarrow$ bread, cookie	1	50%

6.1.2 Basic Algorithms for Association Rule Mining

6.1.2.1 Apriori

The first algorithm was introduced by Agrawal et al. [1] to address the association rule mining issue. The same authors introduced another algorithm named Apriori in their later paper [4] by introducing the monotonicity property of the association rules to improve the performance. Mannila et al. [39] presented an independent work with a similar idea.

Apriori applies a two-stage approach to discover frequent itemsets and confident association rules.

- **Frequent Itemset Discovery.** To find all frequent itemsets, Apriori introduces a candidate generation and test strategy. The basic idea is that it first generates the candidate k-itemsets (i.e., k is 1 at the beginning and is incrementd by 1 in the next cycle), then these candidates will be evaluated whether frequent or not.

Specifically, the algorithm first scans the dataset and the frequent 1-itemsets are found. To discover those frequent 2-itemsets, Apriori generates candidate 2-itemsets by joining 1-itemsets. These candidates are evaluated by scanning the original dataset again. In a similar way, all frequent $(k+1)$-itemsets can be found based on already known frequent k-itemsets.

To improve the performance by avoiding to generate too many yet unnecessary candidates, Apriori introduced a monotonicity property that a $(k+1)$-itemset becomes a candidate only if all its k-subset are frequent. As demonstrated by the authors [4] and many later works, this simple yet efficient strategy largely reduces the candidates to be evaluated.

The frequent itemset mining of the Apriori algorithm is presented in Algorithm 1. The algorithm is executed in a breadth-first search manner. To generate the candidate itemsets with length $k+1$, two k-itemsets with the same $(k-1)$-prefix are joined together (lines 12–13). The joined itemset can be inserted into C_{k+1} only if all its k-subsets are frequent (line 14).

To test the candidate k-itemsets (i.e., count their supports), the database is scanned sequentially and all the candidate itemsets are tested whether they are included in the transaction scanned. By this way, the corresponding support is accumulated (lines 5–9). Finally, frequent itemsets are collected (line 10).

- **Association Rule Mining.** After discovering the frequent itemsets, we can find the frequent and confident association rules straightforward. The approach is similar to the frequent itemset mining algorithm. Because the cost of finding frequent itemsets is high and accounts for most of the whole performance on discovering associate rules, almost all the researches so far have been focused on the frequent itemset generation step.

6.1.2.2 Eclat

Many algorithms had been proposed based on Apriroi idea, in which Eclat [64, 61] is distinct in that it is the first to proposed to generate all frequent itemsets in a depth-first manner, while employing the vertical database layout and uses the intersection based approach to compute the support of an itemset.

Figure 6.1.1 illustrates the key idea of Eclat on candidate support counting. While first scanning of the dataset, it converts the original format (i.e., Table 6.1) into vertical TID list format, as shown in Fig. 6.1.1. For example, the TID list of itemset {*bread*} is {10, 20, 50}, which indicate the transactions that the itemset exist in the original dataset.

To count the support of k-candidate itemset, the algorithm intersects its two $(k-1)$-subset to get the result. For example, as shown in Fig. 6.1.1, to count the support of the itemset {*bread, chocolate*}, it intersects the TID lists of {*bread*} and {*chocolate*}, resulting in {20}. The support is therefore 1.

Algorithm 1: Apriori—Frequent Itemset Mining [4]

Input: A transaction database D, a user specified threshold min_{sup}
Output: Frequent itemsets $\mathcal{F}$
1 $C_1 = \{$1-itemsets$\}$;
2 $k=1$;
3 **while** $C_k \neq NULL$ **do**
4 $\quad$ // Test candidate itemsets
5 $\quad$ **for** *transaction* $T \in D$ **do**
6 $\quad\quad$ **for** *candidate itemsets* $X \in C_k$ **do**
7 $\quad\quad\quad$ **if** $X \subseteq T$ **then** X.support++;
8 $\quad\quad$ **end**
9 $\quad$ **end**
10 $\quad$ $\mathcal{F}_k = \mathcal{F}_k \cup X$, where $X.support \geq min_{sup}$;
11 $\quad$ // Generate candidate itemsets
12 $\quad$ **for** *all* $\{i_1, \ldots i_{k-1}, i_k\}, \{i_1, \ldots i_{k-1}, i'_k\} \in \mathcal{F}_k$ *such that* $i_k < i'_k$ **do**
13 $\quad\quad$ $c = \{i_1, \ldots i_{l-1}, i_k, i'_k\}$;
14 $\quad\quad$ **if** *all k-subsets of c are frequent* **then** $C_{k+1} = C_{k+1} \cup c$;
15 $\quad$ **end**
16 $\quad$ k++;
17 **end**

To reduce the memory used to count the support, Eclat proposed to traverse the lattice (as shown in Fig. 6.1.1) in a depth-first manner. The pseudo code of the Eclat algorithm is presented in Algorithm 2.

The algorithm generates the frequent itemsets by intersecting the tid-lists of all distinct pairs of atoms and evaluating the support of the candidates based on the resulting tid-list (lines 5–6). It calls recursively the procedure with those found frequent itemsets at the current level (line 7–10). This process terminates when all frequent itemsets have been traversed. To save the memory usage, after all frequent itemsets for the next level have been generated, the itemsets at the current level can be deleted.

6.1.2.3 FP-growth

Han et al. [24] proposed a new strategy that mines the complete set of frequent itemsets based on a tree-like structure (i.e., FP-tree). The algorithm applies the divide and conquer approach.

FP-tree construction: FP-tree is constructed as follows [24]: Create the root node of the FP-tree, labeled with "null". Then scan the database and obtain a list of frequent items in which items are ordered with regard to their supports in a descending order. Based on this order, the items in each transaction of the database are reordered. Note that each node n in the FP-tree represents a unique itemset X, i.e., scanning itemset X in transactions can be seen as traversing in the FP-tree from the root to n. All the nodes except the root store a counter which keeps the number of transactions that share the node.

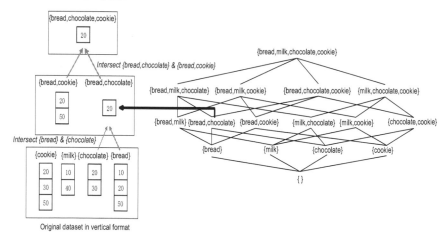

Figure 6.1.1: Eclat mining process (vertical dataset, support count via intersection) [64]

To construct the FP-tree, the algorithm scans the items in each transaction, one at a time, while searching the already existing nodes in FP-tree. If a representative node exists, then the counter of the node is incremented by 1. Otherwise, a new node is created. Additionally, an item header table is built so that each item points to its occurrences in the tree via a chain of node-links. Each item in this header table also stores its support.

Frequent Itemset Mining (FP-growth): To obtain all frequent itemset, Han et al. [24] proposed a pattern growth approach by traversing in the FP-tree, which retains all the itemset association information. The FP-tree is mined by starting from each frequent length-1 pattern (as an initial suffix pattern), constructing its conditional pattern base (a sub-database, which consists of the set of prefix paths in the FP-tree co-occurring with the suffix pattern), then constructing its conditional FP-tree and performing mining recursively on such a tree. The pattern growth is achieved by the concatenation of the suffix pattern with the frequent patterns generated from a conditional FP-tree.

Example 2. Let our example database be the database shown in Table 6.4 with $min_{sup}=2$. First, the supports of all items are accumulated and all infrequent items are removed from the database. The items in the transactions are reordered according to the support in descending order, resulting in the transformed database shown in Table 6.4. The FP-tree for this database is shown in Fig. 6.1.2. The pseudo code of the FP-growth algorithm is presented in Algorithm 3 [24].

Although the authors of the FP-growth algorithm [24] claim that their algorithm does not generate any candidate itemsets, some works (e.g., [20]) have shown that the algorithm actually generates many candidate itemsets since it essentially uses the same candidate generation technique as is used in Apriori but without its prune step. Another issue of FP-tree is that the construction of the frequent pattern tree is a time consuming activity.

Algorithm 2: Eclat—Frequent Itemset Mining [64]

Input: A transaction database D, a user specified threshold min_{sup}, a set of atoms of a sublattice S

Output: Frequent itemsets $\mathcal{F}$

1 **Procedure Elat(S):**
2 **for** *all atoms* $A_i \in S$ **do**
3 $\quad$ $T_i = \emptyset$;
4 $\quad$ **for** *all atoms* $A_j \in S$, *with* $j > i$ **do**
5 $\quad\quad$ $R = A_i \cup A_j$;
6 $\quad\quad$ $\mathcal{L}(\mathcal{R}) = \mathcal{L}(A_i) \cap \mathcal{L}(A_j)$;
7 $\quad\quad$ **if** $support(R) \geq min_{sup}$ **then**
8 $\quad\quad\quad$ $T_i = T_i \cup \{R\}$;
9 $\quad\quad\quad$ $\mathcal{F}_{|\mathcal{R}|} = \mathcal{F}_{|\mathcal{R}|} \cup \{R\}$;
10 $\quad\quad$ **end**
11 $\quad$ **end**
12 **end**
13 **for** *all* $T_i \neq \emptyset$ **do** Eclat(T_i);

Table 6.4: An example database for FP-growth

Tid	Transaction	Ordered Transaction
10	{a, b, d, e, f}	{b, d, f, a, e}
20	{b, f, g}	{b, f, g}
30	{d, g, h, i}	{d, g}
40	{a, c, e, g, j}	{g, a, e}
50	{b, d, f}	{b, d, f}

6.2 Sequential Pattern Mining

The sequential mining problem was first introduced in [5]; two sequential patterns examples are: "80% of the people who buy a television also buy a video camera within a day", and "Every time Microsoft stock drops by 5%, then IBM stock will also drop by at least 4% within three days". The above patterns can be used to determine the efficient use of shelf space for customer convenience, or to properly plan the next step during an economic crisis. Sequential pattern mining is also very important for analyzing biological data [8] [17], in which a very small alphabet (i.e., 4 for DNA sequences and 20 for protein sequences) and long patterns with a typical length of few hundreds or even thousands, frequently appear.

Sequence discovery can be thought of as essentially an association discovery over a temporal database. While association rules [3, 30] discern only intra-event patterns (itemsets), sequential pattern mining discerns inter-event patterns (sequences). There are many other important tasks related to association rule mining, such as correlations [10], causality [46], episodes [38], multi-dimensional patterns [33, 29], max-patterns [9], partial periodicity [23], and emerging patterns [16]. Incisive exploration of sequential pattern mining issue will definitely help to get the efficient solutions to the other research problems shown above.

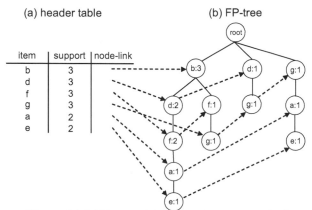

Figure 6.1.2: FP-tree of the example database [24]

Efficient sequential pattern mining methodologies have been studied extensively in many related problems, including the general sequential pattern mining [5, 47, 62, 44, 7], constraint based sequential pattern mining [19], incremental sequential pattern mining [42], frequent episode mining [37], approximate sequential pattern mining [31], partial periodic pattern mining [23], temporal pattern mining in data stream [48], maximal and closed sequential pattern mining [34, 56, 49]. In this section, due to space limitation, we focus on introducing the general sequential pattern mining algorithm, which is the most basic one because all the others can benefit from the strategies it employs, i.e., Apriori heuristic and projection-based pattern growth. More detail and survey on sequential pattern mining can be found in [51, 35].

6.2.1 Sequential Pattern Mining Problem

Let $I = \{i_1, i_2, \ldots, i_k\}$ be a set of items. A subset of I is called an *itemset* or an *element*. A *sequence*, s, is denoted as $\langle t_1, t_2, \ldots, t_l \rangle$, where t_j is an itemset, i.e., $(t_j \subseteq I)$ for $1 \leq j \leq l$. The *itemset*, t_j, is denoted as $(x_1 x_2 \ldots x_m)$, where x_k is an item, i.e., $x_k \in I$ for $1 \leq k \leq m$. For brevity, the brackets are omitted if an *itemset* has only one item. That is, *itemset* (x) is written as x. The number of items in a sequence is called the *length* of the sequence. A sequence with length l is called an *l-sequence*. A sequence, $s_a = \langle a_1, a_2, \ldots, a_n \rangle$, is contained in another sequence, $s_b = \langle b_1, b_2, \ldots, b_m \rangle$, if there exists integers $1 \leq i_1 < i_2 < \ldots < i_n \leq m$, such that $a_1 \subseteq b_{i_1}, a_2 \subseteq b_{i_2}, \ldots, a_n \subseteq b_{i_n}$. We denote s_a a *subsequence* of s_b, and s_b a *supersequence* of s_a. Given a sequence $s = \langle s_1, s_2, \ldots, s_l \rangle$, and an item α, $s \lozenge \alpha$ denotes that s concatenates with α, which has two possible forms, such as *Itemset Extension (IE)*, $s \lozenge \alpha = \langle s_1, s_2, \ldots, s_l \cup \{\alpha\} \rangle$, or *Sequence Extension (SE)*, $s \lozenge \alpha = \langle s_1, s_2, \ldots, s_l, \{\alpha\} \rangle$. If $s' = p \lozenge s$, then p is a *prefix* of s' and s is a *suffix* of s'.

A *sequence database*, S, is a set of tuples $\langle sid, s \rangle$, where sid is a sequence_id and s is a sequence. A tuple $\langle sid, s \rangle$ is said to contain a sequence β, if β is a *subsequence* of s. The support of a sequence, β, in a sequence database, S, is the number of tuples in the database containing β, denoted as $support(\beta)$. Given a user specified positive integer, ε, a sequence, β, is called a frequent sequential pattern if $support(\beta) \geq \varepsilon$.

Algorithm 3: FP-growth [24]

Input: A transaction database D, a frequent pattern tree FP-tree, a user specified
  threshold min_{sup}
Output: Frequent itemsets $\mathcal{F}$
1 **Method:** *call* **FP-growth**(FP-tree, null)
2 **Procedure FP-growth**($Tree$, α):
3 **if** $Tree$ *contains a single prefix-path* **then**
4 | Let P be the single prefix-path part of $Tree$;
5 | Let Q be the multipath part with the top branching node replaced by a *null* root;
6 | **for** *each combination β of the nodes in P* **do**
7 | | Generate pattern $\beta \cup \alpha$ with *support*=minimum support of nodes in β;
8 | | Let *freq_pattern_set*(P) be the set of patterns generated;
9 | **end**
10 **else**
11 | Let Q be $Tree$;
12 **end**
13 **for** *each item $a_i \in Q$* **do**
14 | generate pattern $\beta = a_i \cup \alpha$ with *support*=$a_i.support$;
15 | construct β's conditional pattern-base and then β' conditional FT-tree $Tree_\beta$;
16 | **if** $Tree_\beta \neq \emptyset$ **then** call Fp-growth($Tree_\beta, \beta$);
17 | Let *freq_pattern_set*(Q) be the set of patterns generated;
18 **end**
19 return($freq\_pattern\_set(P) \cup freq\_pattern\_set(Q) \cup (freq\_pattern\_set(P) \times freq\_pattern\_set(Q)))$;

6.2.2 Existing Sequential Pattern Mining Algorithms

There are many sequential pattern mining algorithms introduced, which can be classified into three groups [36]. One group is Apriori-like algorithm, such as Apriori-all [5], GSP [47], SPADE [62], and SPAM [7], the second group is projection-based pattern growth, such as PrefixSpan [44], the third group is early prune based strategy, such as LAPIN [58, 60].

6.2.2.1 AprioriALL

Sequential pattern mining was first introduced in [5] by Agrawal, an Apriori based algorithm, i.e., AprioriALL, was proposed. Given the transaction database as illustrated in Fig. 6.2.1, the mining process can be implemented in five steps:

- **Sort Step:** The database is sorted according to the customer ID and the transaction time, as illustrated in Fig. 6.2.1.
- **L-itemsets Step:** The sorted data is first scanned to obtain those frequent (or *large*) 1-itemsets based on the user specified support

CID	Transaction Time	Items
10	Sep. 5, 2011	bread
10	Sep. 9, 2011	cookie
10	Sep. 10, 2011	banana, cookie
10	Sep. 12, 2011	chocolate
10	Sep. 20, 2011	bread, milk, cookie
10	Sep. 23, 2011	bread
10	Sep. 26 2011	chocolate
20	Sep. 7, 2011	milk
20	Sep. 11, 2011	cookie, chocolate
20	Sep. 13, 2011	bread
20	Sep. 16, 2011	cookie
20	Sep. 22, 2011	milk, chocolate
30	Sep. 6, 2011	chocolate
30	Sep. 9, 2011	milk, cookie
30	Sep. 11, 2011	bread, cookie
30	Sep. 15, 2011	cookie, chocolate

Figure 6.2.1: Database Sorted by Customer ID and Transaction Time

Large Itemsets	Mapped To
apple	a
banana	b
strawberry	c
pear	d

Figure 6.2.2: Large Itemsets

threshold. Suppose the minimal support is 70%, in this case the minimal support count is 2, the result of large 1-itemsets is listed in Fig. 6.2.2.

- **Transformation Step:** We map the large itemsets into a series of integers and the original database is converted by replacing the itemsets. For example, with the help of the mapping table in Fig. 6.2.2, the transformed database is obtained, as shown in Fig. 6.2.3.
- **Sequence Step:** The transformed database is scanned and mined to find all the frequent patterns.
- **Maximal Step:** We remove those patterns which are contained in other sequential patterns. In other words, only maximal sequential patterns remain.

Customer ID	Customer Sequence
10	$<ac(bc)d(abc)ad>$
20	$<b(cd)ac(bd)>$
30	$<d(bc)(ac)(cd)>$

Figure 6.2.3: Transformed Database

Table 6.1: AprioriAll Candidate Generation L_3 to C_4 [5]

Large 4-sequences	Candidate 5-sequences
$\langle b(ac)d \rangle$	$\langle (bc)(ac)d \rangle$
$\langle bcad \rangle$	$\langle d(bc)ad \rangle$
$\langle bdad \rangle$	$\langle d(bc)da \rangle$
$\langle bdcd \rangle$	$\langle d(bc)(ad) \rangle$
$\langle (bc)ad \rangle$	
$\langle (bc)(ac) \rangle$	
$\langle (bc)cd \rangle$	
$\langle c(ac)d \rangle$	
$\langle d(ac)d \rangle$	
$\langle dbad \rangle$	
$\langle d(bc)a \rangle$	
$\langle d(bc)d \rangle$	
$\langle dcad \rangle$	

Among all these steps, the sequence step is the most time consuming one and therefore, researchers focused on this step. AprioriAll [5] was first proposed based on the Apriori algorithm in association rule mining [3]. Two phases are utilized to mine sequential patterns, i.e., candidate generation and test.

The phase for generating candidates is similar to the AprioriGen in [3]. The Apriori property is applied to prune those candidate sequences whose subsequence is not frequent. The difference is that when the authors generate the candidate by joining the frequent patterns in the previous pass, different order of combination make different candidates. For example: from the items, a and b, three candidates $\langle ab \rangle$, $\langle ba \rangle$ and $\langle (ab) \rangle$ can be generated. But in association rule mining only $\langle (ab) \rangle$ is generated. The reason is that in association rule mining, the time order is not taken into account. Obviously the number of candidate sequences in sequential pattern mining are much larger than the size of the candidate itemsets in association rule mining during the generation of candidate sequences. Table 6.1 shows how to generate candidate 5-sequences by joining large 4-sequences. By scanning the large 4-itemsets, it finds that the first itemsets $\langle (bc)ad \rangle$ and second itemsets $\langle (bc)(ac) \rangle$ share their first three items, according to the join condition of Apriori they are joined to produce the candidate sequence $\langle (bc)(ac)d \rangle$. Similarly other candidate 5-sequences are generated.

For the second phase, i.e., test phase, is simple and straightforward. The database is scanned to count the supports of those candidate sequences. As a result, the frequent sequential patterns can be found.

Due to the efficiency and simplicity of the AprioriAll algorithm, which is the first algorithm on mining sequential patterns, the core idea of AprioriAll is applied by many other algorithms. The problems of AprioriAll are that there are many candidates generated and multiple passes over the databases are very time consuming.

6.2.2.1.1 GSP

Srikant and Agrawal generalized the definition of sequential pattern mining problem in [47] by incorporating some new properties, i.e., time constraints, transaction relaxation, and *taxonomy*. For the time constraints, the *maximum gap* and the *minimal gap* are defined to specified the gap between any two adjacent transactions in the sequence. When testing a candidate, if any gap of the candidate falls out of the range between the maximum gap and the minimal gap, then the candidate is not a pattern. Furthermore, the authors relaxed the definition of transaction by using a *sliding window*, that when the time range between two items is smaller than the sliding window, these two items are considered to be in the same transaction. The taxonomy is used to generate multiple level sequential patterns.

In [47], the authors proposed a new algorithm which is named GSP to efficiently find the generalized sequential patterns. Similar to the AprioriAll algorithm, there are two phases in GSP, i.e., candidate generation and test.

In the candidate generation process, the candidate k-sequences are generated based on the frequent $(k-1)$-sequences. Given a sequence $s = \langle s_1, s_2, \ldots, s_n \rangle$ and subsequence c, c is a contiguous subsequence of s if any of the following conditions holds: (1) c is derived from s by dropping an item from either s_1 or s_n; (2) c is derived from s by dropping an item from an element s_j that has at least 2 items; and (3) c is a contiguous subsequence of $\hat{c}$, and $\hat{c}$ is a contiguous subsequence of s. Specifically, the candidates are generated in two phases:

- **Joint Phase:** Candidate k-sequences are generated by joining two $(k-1)$-sequences that have the same contiguous subsequences. When we join the two sequences, the item can be inserted as a part of the element or as a separated element. For example, because $\langle d(bc)a \rangle$ and $\langle d(bc)d \rangle$ have the same contiguous subsequence $\langle d(bc) \rangle$, then we know that candidate 5-sequence $\langle d(bc)(ad) \rangle$, $\langle d(bc)ad \rangle$ and $\langle d(bc)da \rangle$ can be generated.
- **Prune Phase:** The algorithm removes the candidate sequences which have a contiguous subsequence whose support count is less than the minimal support. Moreover, it uses a hash-tree structure [41] to reduce the number of candidates.

The process for generating candidates in the example database is shown in Fig. 6.2. For GSP, the difficulty is that the support of candidate sequences is not easy to count due to the introduced generalization rules, while this is not a problem for AprioriAll. GSP devises an efficient strategy which includes two phases, i.e., forward and backward phases (which are repeated until all the elements are found): (1) **Forward Phase:** It looks for successive elements of s in d, as long as the difference between the end-time of the element and the start-time of the previous element is less than the maximum gap. If the difference is greater than the maximum gap, it switches to the backward phase. If an element is not found, then s is not contained in d; (2) **Backward Phase:** It tries to pull up the previous element. Suppose s_i is the current element and end-time$(s_i)=t$. It checks whether there are some transactions containing s_{i-1} and the corresponding transaction-times are larger than the maximum gap. Since after pulling up s_{i-1}, the difference between s_{i-1} and s_{i-2} may not satisfy the gap constraints, the backward pulls back until the difference of s_{i-1} and s_{i-2} satisfies the maximum gap or the first element has been pulled up. Then the algorithm switches to the forward phase. If all the elements can not be pulled up, then d does not contain s.

Table 6.2: GSP Candidate Generation L_4 to C_5 [47]

Large 4-sequences	Candidate 5-sequences after joining	Candidate 5-sequences after pruning
⟨b(ac)d⟩	⟨(bc)(ac)d⟩	⟨(bc)(ac)d⟩
⟨bcad⟩	⟨d(bc)ad⟩	⟨d(bc)ad⟩
⟨bdad⟩	⟨d(bc)da⟩	
⟨bdcd⟩	⟨d(bc)(ad)⟩	
⟨(bc)ad⟩		
⟨(bc)(ac)⟩		
⟨(bc)cd⟩		
⟨c(ac)d⟩		
⟨d(ac)d⟩		
⟨dbad⟩		
⟨d(bc)a⟩		
⟨d(bc)d⟩		
⟨dcad⟩		

For generalized rule, the authors [47] introduced taxonomy knowledge into the mining process. The taxonomies are incorporated by extending sequences with corresponding taxonomies. The original sequences are therefore, replaced by their extended versions. As a result, the number of rules becomes larger because the sequences become more dense and redundant rules are produced. To avoid uninteresting rules, the ancestors are firstly precomputed for each item and those are not in the candidates are removed. Moreover, the algorithm does not count the sequential patterns that contain both the item and its ancestors. In a summary, the generalized sequential patterns take more attributes into account and thus, can be applied to real applications easily.

6.2.2.1.2 SPADE

Zaki introduced another efficient algorithm, i.e., SPADE [62], to find frequent sequences using efficient lattice search techniques and simple joins. To discover all the patterns, SPADE needs to scan the database three times. It divides the mining problem into smaller ones to conquer and at the same time makes it possible that all the necessary data is located in memory. The core idea of SPADE, is devised based on that of Eclat [64], one of the efficient algorithms for association rule mining. From the extensive experimental evaluation [62], we can see that SPADE is very efficient in finding sequential patterns.

The mining process of SPADE can be illustrated through a concrete example. Firstly, the sequential database is transformed into a vertical format, i.e., id-list database, in which each id is associated with its corresponding customer sequence and transaction. The vertical version of the original database (as shown in Fig. 6.2.1) is illustrated in Fig. 6.2.4. For example, we know that the id-list of item a is (100, 1), (100, 5), (100, 6), (200, 3), and (300, 3), where each pair (*SID*:*TID*) indicates the specific sequence and transaction that *a* locates. By scanning the vertical database, frequent 1-sequences can be easily obtained. To find the frequent 2- sequences, the original database is scanned again and the new vertical to horizontal database is constructed by grouping those items with SID and in increase order of TID, which is shown in Fig. 6.2.5. By scanning the database 2-length patterns can be discovered. A lattice is constructed based on these 2-length patterns, and the lattice can be further decomposed into different classes, where those patterns that have the same prefix belong to the same class. Such kind of decomposition make it possible that the partitions are small enough to be loaded into the memory. SPADE then applies temporal joins to find all other longer patterns by enumerating the lattice [62].

a		b		c		d	
SID	TID	SID	TID	SID	TID	SID	TID
10	1	10	3	10	2	10	4
10	5	10	5	10	3	10	7
10	6	20	1	10	5	20	2
20	3	20	5	20	2	20	5
30	3	30	2	20	4	30	1
				30	2	30	4
				30	3		
				30	4		

Figure 6.2.4: Vertical id-List

SID	(Item, TID) pairs
10	(a, 1) (c, 2) (b, 2) (c, 2) (d, 4) (a, 5) (b, 5) (c, 5) (a, 6) (d, 7)
20	(b, 1) (c, 2) (d, 2) (a, 3) (c, 4) (b, 5) (d, 5)
30	(d, 1) (b, 2) (c, 2) (a, 3) (c, 3) (c, 4) (d, 4)

Figure 6.2.5: Vertical to Horizontal Database

In SPADE, two strategies are introduced to traverse all the candidate sequences, i.e., *breadth first search* (BFS) and *depth first search* (DFS). For the first strategy, i.e., BFS, the candidate sequences are generated in a recursive bottom up manner. For instance, to generate the 3-length patterns, all the 2-length patterns have to be obtained. On the contrary, for the second strategy, i.e., DFS, it only requires that one 2-length pattern and one k-length pattern to generate a $(k+1)$-length sequence (assume that the last item of the k-pattern is the same as the first item of the 2-pattern). Therefore, there is always a trade-off between BFS and DFS: while BFS needs more memory to store all the consecutive 2-length patterns, it has the advantage that more information is obtained to prune the candidate k-sequences. All the k-length patterns are discovered by temporal or equality joining the frequent $(k-1)$-length patterns which have the same $(k-2)$-length prefix. To furthermore improve the efficiency, SPADE applies the commonly used Apriori strategy.

To explain in detail the temporal join process of SPADE, we use a concrete example as shown in Fig. 6.2.6. After the 1-length patterns, i.e., a and b, are obtained, to join these two patterns, we can test the three candidate sequences, ab, ba and (ab). The joining operation is indeed to compare the SID, TID pairs of the two $(k-1)$-length patterns. For example, the pattern b has two pairs {100, 3}, {100, 5} which are larger than (behind) the pattern a's one pair ({100, 1}), in the same customer sequence. Hence, $\langle ab \rangle$ should exist in the same sequence. The other candidate sequences' support can be accumulated in a similar way. Figure 6.2.6 shows the process.

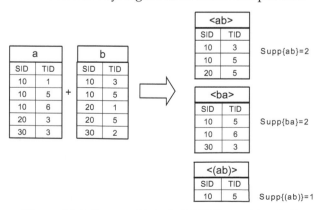

Figure 6.2.6: Temporal join in SPADE algorithm [62]

6.2.2.1.3 SPAM

The SAPM algorithm [7] was introduced based on the key idea of SPADE. The difference is that SPAM applies a bitmap representation of the database instead of {*SID, TID*} pairs used in the SPADE algorithm. Therefore by using bitwise operations SPAM can obtain a better performance than SPADE and others on longer large databases.

The mining process of SPAM can be explained as follows. When we scan the database for the first time, a vertical bitmap is constructed for each item in the database, and each bitmap has a bit corresponding to each itemset (element) of the sequences in the database. If an item appears in an itemset, the bit corresponding to the itemset of the bitmap for the item is set to one; otherwise, the bit is set to zero. The size of a sequence is the number of itemsets contained in the sequence. Figure 6.2.7 shows the bitmap vertical table of that in Fig. 6.2.3. A sequence in the database of size between 2^k+1 and 2^{k+1} is considered as a 2^{k+1}-bit sequence. The bitmap of a sequence will be constructed according to the bitmaps of items contained in it.

To generate and test the candidate sequences, SPAM uses two steps, S-step and I-step, based on the lattice concept. As a depth-first approach, the overall process starts from S-step and then I-step. To extend a sequence, the S-step appends an item to it as the new last element, and the I-step appends the item to its last element if possible. Each bitmap partition of a sequence to be extended is transformed first in the S-step, such that all bits after the first bit with value one are set to one. Then the resultant

SID	TID	{a}	{b}	{c}	{d}
10	1	1	0	0	0
10	2	0	0	1	0
10	3	0	1	1	0
10	4	0	0	0	1
10	5	1	1	1	0
10	6	1	0	0	0
10	7	0	0	0	1
20	1	0	1	0	0
20	2	0	0	1	1
20	3	1	0	0	0
20	4	0	0	1	0
20	5	0	1	0	1
30	1	0	0	0	1
30	2	0	1	1	0
30	3	1	0	1	0
30	4	0	0	1	1

Figure 6.2.7: Bitmap Vertical Table

bitmap of the S-step can be obtained by doing ANDing operation for the transformed bitmap and the bitmap of the appended item. Figure 6.2.8 illustrates how to join two 1-length patterns, a and b, based on the example database in Fig. 6.2.3. On the other hand, the I-step just uses the bitmaps of the sequence and the appended item to do ANDing operation to get the resultant bitmap, which extend the pattern $\langle ab \rangle$ to the candidate $\langle a(bc) \rangle$. The support counting becomes a simple check how many bitmap partitions not containing all zeros.

The main drawback of SPAM is the huge memory consumption. For example, although an item, a, does not exist in a sequence, s, SPAM still uses one bit to represent the existence of a in s. This disadvantage restricts SPAM as a best algorithm on mining large datasets in limit resource environments.

6.2.2.1.4 PrefixSpan

Pei et al. introduced the PrefixSpan algorithm in [43]. The key idea of the PrefixSpan algorithm is to apply database projection to make the database smaller for next iteration and thus, improve the performance. The authors claimed that in PrefixSpan there is no need for candidates generation [43].[1] It recursively projects the database by already found short length patterns. Different projection methods were introduced, i.e., *level-by-level* projection, *bi-level* projection, and *pseudo* projection.

The workflow of PrefixSpan is presented as follows. Assume that items within transactions are sorted in alphabetical order (it does not affect the result of discovered patterns). Similar to other algorithms, the first step of PrefixSpan is to scan the database to get the 1-length patterns. Then the original database is projected into different partitions with regard to the frequent 1-length pattern by taking the corresponding pattern as the prefix. For example, Fig. 6.2.9 (b) shows the projected databases with the frequent (or large) 1-length patterns as their prefixes. The next step is to scan the projected database of γ, where γ could be any one of the 1-length patterns. After the scanning, we can obtain the frequent 1-length patterns in the projected database. These patterns, combined with their common prefix γ, are deemed as 2-length patterns. The process will be executed recursively, that the projected database is partitioned by the k-length patterns, to find those $(k+1)$-length patterns, until the projected database is empty or no more frequent patterns can be found.

---

[1]However, some works (e.g., [58, 60]) have found that PrefixSpan also needs to test the candidates, which are existing in the projected database.

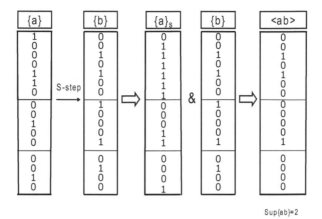

Sup{ab}=2

Figure 6.2.8: SPAM S-Step join [7]

The introduced strategy is named *level-by-level* projection. The main computation cost is the time and space usage when constructing and scanning the projected databases, as shown in Fig. 6.2.9 (b). To improve the efficiency, another strategy named *bi-level* projection was proposed to reduce the cost of building and scanning the projected databases [44]. The difference between the two projection strategies is that, in the second step of bi-level projection, a $n \times n$ triangle matrix (called S-matrix) is constructed by scanning the database again, as shown in Fig. 6.2.9 (c). This matrix represents all the supports of 2-length candidates. For example, $M[\langle d \rangle, \langle a \rangle]=(3, 3, 0)$ indicates that the supports of $\langle da \rangle$, $\langle ad \rangle$, and $\langle (ad) \rangle$ are 3, 3, and 0, respectively. The original database is then projected with regard to the frequent 2-length patterns in the S-matrix and the projected databases are scanned, respectively. The process recursively follows such a projection and scanning manner to find all the patterns. This strategy, however, seems to be not always optimal, as stated in [44].

A further optimization named *pseudo projection* was proposed in [43] to make the projection more efficient when the projected database can be loaded into the memory. The strategy is fulfilled by employing a pair of pointer and offset to indicate the position of each projection database instead of copying the data each time. The drawback is that the size of the (projected) database can not be too large.

In a brief summary, PrefixSpan improves the performance of mining sequential patterns by using database projection, that it scans smaller projected databases in each iteration. The main problem of PrefixSpan, however, is that it is time consuming on scanning the projected database, which may be very large if the original dataset is huge.

Large Itemsets	Projected Database
a	<c(bc)d(abc)ad> <c(bd)> <(_c)(cd)>
b	<(_c)d(abc)ad> <(cd)ac(bd)> <(_c)(ac)(cd)>
c	<c(bc)d(abc)ad> <(_d)ac(bd)> <(ac)(cd)>
d	<(abc)ad> <ac(bd)> <(bc)(ac)(cd)>

Customer ID	Customer Sequence
10	<ac(bc)d(abc)ad>
20	<b(cd)ac(bd)>
30	<d(bc)(ac)(cd)>

(a) Example Database   (b) Projected Database

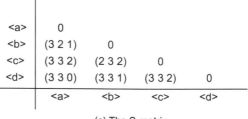

<a>	0			
<b>	(3 2 1)	0		
<c>	(3 3 2)	(2 3 2)	0	
<d>	(3 3 0)	(3 3 1)	(3 3 2)	0
	<a>	<b>	<c>	<d>

(c) The S-matrix

Figure 6.2.9: PrefixSpan Mining Process [43]

6.2.2.1.5 LAPIN

LAPIN was proposed in [58, 60]. The basic idea of LAPIN is that the last position of each item is important and can be utilized to improve the performance to count the frequency of the candidates. The main difference between LAPIN and previous significant algorithms is the scope of the search space. PrefixSpan scans the whole projected database to find the frequent patterns. SPADE temporally joins the whole id-list of the candidates to get the frequent patterns of next layer. LAPIN can get the same results by scanning only part of the search space of PrefixSpan and SPADE, which are indeed the last positions of the items.

For the same example shown in Fig. 6.2.9, LAPIN constructs the item *a*'s projected last position lists when testing the candidates whose prefix is *a*, as illustrated in Table 6.3. We can see that to obtain the 2-length patterns whose prefix is *a*, LAPIN only needs to scan 8 elements while PrefixSpan needs to test 16 times. The reason is that redundant testing the same item in the projected database is useless for contribution of counting of the candidates. From the example, we can know that LAPIN is a prefix growth algorithm with efficient pruning strategy. It employs a depth first search of the lexicographic tree to grow the sequences with the help of the projected item last position databases.

In addition, the same authors introduced some variant versions of LAPIN, i.e., LAPINSPAM [57] and LAPIN-Web [59]. The first one is devised based on the SPAM algorithm [7] which utilizes bit information to further improve the efficiency. The latter one, i.e., LAPIN-Web [59], is introduced to specifically extract the users' frequent access patterns with regard to the log data.

Table 6.3: Item A's Projected Last Position Lists

Customer ID	Projected Last Position Lists
10	$c_{last} = 5, a_{last} = 6, d_{last} = 7$
20	$c_{last} = 5, b_{last} = 6, d_{last} = 6$
30	$c_{last} = 4, d_{last} = 4$

As clarified in [36], the LAPIN strategy can be deemed as one of the promising techniques in the sequential pattern mining literature.

6.3 Frequent Subtree Mining

Frequent subtree mining could be seen as an extension issue of frequent itemset and sequence mining because the data structure of the former is more complex than that of the latter. There are many applications based on the frequent tree mining, such as Web mining, bioinformatics, computer networks, and so forth. In this section, we will introduce the basic concepts and algorithms for mining frequent subtrees. In essential, most of these algorithms follow the same spirit of the techniques developed in frequent itemset mining. More detail and survey on frequent subtree mining can be found in [11].

6.3.1 Frequent Subtree Mining Problem

The frequent subtree mining problem is defined as follows [11]. Given a class of trees T, a threshold min_{sup}, a transitive subtree relation $P \preceq Q$ between trees $P, Q \in T$, a finite data set of trees $D \subseteq T$, the frequent tree mining problem is the problem of finding all trees $P \subseteq T$ such that no two trees in P are isomorphic and for all $P \in P$: $freq(P,D) = \sum_{Q \in D} d(P, Q) \geq min_{sup}$, where d is an anti-monotone function such that $\forall Q \in T : d(P', Q) \geq d(P, Q)$ if $P' \preceq P$. The simplest choice for function d is given by the indicator function: $d(P, Q) = 1$, if $P \preceq Q$, otherwisw $d(P, Q)=0$.

For ease of exposition, in this chapter, we only consider the simple case, that the frequency of a pattern tree is defined by the number of trees in the data set that contains the pattern tree. The frequency definition is denoted as *transaction based frequency*, which is similar to that of itemset or sequence frequency. Because of the transitivity property of the subtree relation, the indicator function is anti-monotone and can be utilized to improve the mining efficiency. Based on the definition of the pattern frequency, the support can be defined as follows: $sup(P, D)=freq(P, D)/|D|$, which is also similar to that of itemset or sequence based support.

Following the same strategy as other structure pattern mining algorithms applied, we can utilize the classic method, i.e., the generate and test [1], to discover all the frequent subtrees. The common work flow can be executed the following two steps recursively, where P is set an empty tree firstly: (1) calculate $freq(P)$ by finding all $Q \in D$ with $P \preceq Q$; and (2) let $P=suc(Q)$. Note that $suc(P)$ is a possible approach that determines the successor of P tree. It should guarantee that all the possible trees are enumerated exactly once and only once. There are many possible methods to decide the concrete implementation. They are different at the data structure used and the performance cost.

Algorithm 4: The TreeMiner algorithm [63]

Input: D, σ, $\{T_1^k, \ldots, T_m^k\}$, $\mathcal{F}(\sigma, D, \preceq_e)$
Output: $\mathcal{F}(\sigma, D, \preceq_e)$

1 **for** $i \leftarrow 1$ *to* $i=m$ **do**
2 $\quad$ $F_i^{k+1} \leftarrow \emptyset$;
3 $\quad$ **for** $j \leftarrow 1$ *to* $j=m$ **do**
4 $\quad\quad$ $C_i^{k+1} \leftarrow \emptyset$;
5 $\quad\quad$ $C_i^{k+1} \leftarrow \otimes(T_i^k, T_j^k,)$;
6 $\quad\quad$ **for** *all* $T_{i,j}^{k+1} \in C_{i,j}^{k+1}$ *as* $supp(T_{i,j}^{k+1}) \geq \sigma$ **do**
7 $\quad\quad\quad$ $F_i^{k+1} \leftarrow F_i^{k+1} \cup \{T_{i,j}^{k+1}\}$;
8 $\quad\quad$ **end**
9 $\quad$ **end**
10 $\quad$ $\mathcal{F}(\sigma, D, \preceq_e) \leftarrow \mathcal{F}(\sigma, D, \preceq_e) \cup \{F_i^{k+1}\}$;
11 $\quad$ TreeMiner(D, σ, F_i^{k+1}, $\mathcal{F}(\sigma, D, \preceq_e)$);
12 **end**

6.3.2 Data Structures for Storing Trees

There are many possible data structures can be used for storing trees. For example, the adjacency matrix and the first-child-next-sibling are commonly utilized. In addition to these data structures, some other tree representations have been also introduced for different purposes. For example, to save space, some canonical representations are proposed because they are more compact than the commonly used data structures. Another reason is that because there are always many possibilities to represent the same tree information for labeled trees, using a unique way is important and essential for mining process. An effective representation, therefore, facilitates the comparison process. We will introduce different approaches that were proposed for frequent subtree mining.

6.3.2.1 TreeMiner

Zaki introduced the *TreeMiner* algorithm [63] to mine frequently ordered subtrees. The basic idea is that it applies both breadth first search (BFS) and depth first search (DFS) to traverse the whole search space finding the frequent subtrees. Similar to other structure mining algorithms in the literature, *TreeMiner* also applies the Apriori rule, i.e., all subtrees of a frequent tree are frequent. Moreover, the author introduces an effective strategy that by finding an observation if we remove either one of the last two vertices at the end of the string encoding of a rooted ordered tree, we can obtain the string encoding of a valid embedded subtree. Based on this observation, Zaki proposed to use BFS and DFS integratedly that generates the candidate $(k+1)$-subtrees by joining two frequent k-subtrees which have the same prefix string encodings with $(k-1)$-length. This idea is similar to that of SPADE [62] for sequential pattern mining.

Algorithm 5: The FREQT algorithm [6]

    Input: D, σ, $\{T_1^k, \ldots, T_m^k\}$, $\mathcal{F}(\sigma, D, \preceq_i)$
    Output: $\mathcal{F}(\sigma, D, \preceq_i)$

1 **for** $i \leftarrow 1$ to $i = m$ **do**
2      $F_i^{k+1} \leftarrow \emptyset$;
3      $C_i^{k+1} \leftarrow \emptyset$;
4      $C_i^{k+1} \leftarrow extension(T_i^k, OCL(T_i^k))$;
5   **for** *all* $T_i^{k+1} \in C_i^{k+1}$ *such that* $supp(T_i^{k+1}) \geq \sigma$ **do**
6        $\mid$ $F_i^{k+1} \leftarrow F_i^{k+1} \cup \{T_i^{k+1}\}$;
7   **end**
8      $\mathcal{F}(\sigma, D, \preceq_i) \leftarrow \mathcal{F}(\sigma, D, \preceq_i) \cup \{F_i^{k+1}\}$;
9   FREQT$(D, \sigma, F_i^{k+1}, \mathcal{F}(\sigma, D, \preceq_i))$;
10 **end**

To count the frequency of the candidate subtrees, similar to SPADE [62], *TreeMiner* introduces the vertical format to represent the data. Specifically, the scope of a node is defined as between the preorder number of it and the preorder number of the rightmost node of it. From the definition, we can know that the size of the data tree could be very large and this is an issue for large dataset. The pseudo code of the *TreeMiner* algorithm is illustrated in Fig. 4. Refer [63] for detail.

6.3.2.2 FREQT

Asai et al. proposed the *FREQT* algorithm [6] to find the frequent induced subtrees. The basic idea follows the well known property, i.e., Apriori. To generate the candidates, *FREQT* applies the rightmost extension strategy that, a k-tree is extended to a candidate $(k+1)$-tree by adding a new node to the node at the rightmost branch of the k-tree. By this way, we know that the parent tree can be uniquely determined. This strategy also guarantees that each candidate subtrees are traversed exactly only once. Similar to other structure pattern mining (i.e., itemset or sequence mining), the algorithm starts to find 1-patterns and then grows the pattern by increasing 1 and so forth to find all the frequent subtrees. The mining process is terminated when there is no possible extension can be made. Figure 5 shows the pseudo code of the *FREQT* algorithm.

To extend the frequent k-tree to the candidate $(k+1)$-trees, the *FREQT* algorithm utilizes the rightmost extension strategy. Firstly all the siblings of the nodes on the rightmost path of the K-tree are determined, and then the children of the rightmost leaf can be found. Based on these children nodes, the candidate $(k+1)$-tree can be determined. An intuitive idea to implement this strategy, is that we only scan a small part of the tree, instead of scanning the whole data, to improve efficiency. Some techniques have been introduced in [6] to tackle this issue. It utilizes a list of pointers for each tree, to point to the nodes of the pattern map.

Moreover, only the occurrences of the rightmost leaf of the tree is saved to reduce the space cost.

6.3.2.3 HybridTreeMiner

To further improve the efficiency of frequent subtree mining, Chi et al. [14] has proposed the *HybridTreeMiner* algorithm which, similar to *TreeMiner*, applies both the breadth first search and the depth first search strategies. The basic idea of *HybridTreeMiner* also follows the traditional generate-and-test technique. To efficiently generate the candidates to be tested, the authors introduced the tree representation, i.e., breadth first canonical form, to facilitate traversal of all possible subtree candidates. A disadvantage of the algorithm is that it cannot generate all the candidates in constant time because of the complexity cost.

Similar to *TreeMiner*, the *HybridTreeMiner* algorithm joins two k-trees which have the common prefix $(k-1)$-trees, to generate the candidate subtree. For those trees which cannot be generated by joining, *HybridTreeMiner* borrows the idea of *FREQT*, that extends the frequent subtrees to obtain the larger candidates. There are several effective strategies proposed in [13, 14] to address the issue of tree *authomorphisms* during the joining and extending processes. Moreover, for different types of trees, the authors introduced different approaches to improve the efficiency of the algorithms. For example, to deal with the free trees, the algorithm is extended by utilizing the breadth first tree encoding. By this way, it can take account for a small part of all the rooted trees. Another strategy introduced by *HybridTreeMiner* is that the occurrence lists of the subtrees are proposed and the authors explained how they are joined for generating the candidates in Chi et al. [14].

6.3.2.4 Gaston

Nijssen et al. proposed another algorithm named *Gaston* [40]. Based on the similar idea of *TreeMiner* and *HybridTreeMiner*, the *Gaston* algorithm applies both the breadth first search and depth first search strategies. There are several phases introduced for the whole mining process. Firstly, it extracts the frequent undirected paths by traversing all the possible ways. To facilitate the process, the authors introduced an effective representation for trees which can be built in reasonable time for large data; then it deals with these paths as the start point of a rooted tree, and joins or extends them with rightmost path extension technique to generate the candidates and test.

6.3.3 Maximal and closed frequent subtrees

A main issue for all the previous work is that the resultant frequent subtrees may be very large and it can grow exponentially when the size of the original data increases. As a result, how to efficiently obtain them is important. Moreover, it is very difficult to clarify the whole result because of the huge size of them. To tackle these problems, the maximal and closed frequent subtrees have been introduced [50, 54, 12], which borrows the idea from the literature of itemset and sequence mining. The definition of the maximal frequent subtree is that none of a maximal frequent subtree's super trees are frequent. By this way, the discovered frequent patterns can be reduced dramatically, which facilitates the mining process and the explanation of the results. The basic idea of [50, 54] is that they first find all the frequent subtrees, and then filter out those non-maximal patterns. This technique, although simple to be implemented, is time consuming. To tackle the issue, Chi et al. [12] proposed the CMTreeMiner algorithm, which extracts the maximal patterns without first finding all the frequent ones. Furthermore, the authors also introduced to discover closed frequent patterns. A tree is closed if none of its super trees has the same support. By this way, the result can be analyzed with more meaningful information.

6.4 Frequent Subgraph Mining

Frequent subgraph mining (FSM) is an important issue because it is the basis for many applications, such as web mining, bioinformatics, computer networks, and so forth. Most of the existing frequent subgraph methods follow the similar strategies with that proposed in the traditional frequent itemst mining, i.e., Apriori rule. However, the higher complexity of the former issue introduces some unique properties and thus, special solutions for graphs have been presented in the literature.

The key idea of frequent subgraph mining, similar to other structure pattern mining (i.e., itemset, sequence, tree), is that the generate-and-test strategy is implemented during the mining process. Firstly it generates some candidate subgraphs by applying breadth first search or depth first search; and then it tests these candidates to decide whether these subgraphs occur above some predefined threshold. There are several issues need to be tackled. For example, how the candidate subgraphs are to be generated without duplication and none of them are missing? How to efficiently count the frequency of these candidate subgraphs? To tackle the first issue, the Apriori rule has been commonly utilized. In this section, we will briefly introduce the basic concepts and algorithms for frequent subgraph mining. More detail on this issue can be seen in [28, 15].

6.4.1 Problem Definition

The definition presented in this section follows that of [28, 15]. A subgraph g is deemed as frequent if its occurrence is greater than some predefined threshold. The occurrence of a subgraph can be deemed as its support, which is defined by the number of graph transactions that g exists. No matter how many times g occurs in a graph transaction, it accounts for no greater than one count. Given a database $G = \{G_1, G_2, \ldots, G_T\}$ and a threshold min_{sup}, the set of graph transactions where a subgraph g exists is defined by $\xi_G(g) = \{G_i \mid g \subseteq G_i\}$. As a result, the *support* of g is defined as $sup_G(g) = |\xi_{G(g)}| / T$, where $|\xi_G(g)|$ denotes the size of $\xi_G(g)$ and T is the number of graph transactions. We say that g is frequent if the following holds: $sup_G(g) \geq min_{sup}$.

A labeled graph is denoted as $G(V, E, L_V, L_E, \theta)$, where V is a set of vertices, $E \subseteq V \times V$ is a set of edges, L_V is a set of vertex labels, L_E is a set of edge labels, and θ is a function that maps V to L_V and E to L_E. A *path* is defined as a set of vertices in G which could be ordered that two vertices construct an edge if they are consecutive. If for all the $e \in E$, e is an (un) ordered pair of vertices, then we say that G is *(un)directed*. If there exists a path for every pair of vertices in G, we say that G is *connected*. Otherwise, G is *disconnected*. If G contains no cycle, then way denote G as *acyclic*. If every pair of vertices is connected by an edge, we say that G is *complete*. Given two graphs $G_1(V_1, E_1, L_{V_1}, L_{E_1}, \theta_1)$ and $G_2(V_2, E_2, L_{V_2}, L_{E_2}, \theta_2)$, G_1 is a *subgraph* of G_2, if G_1 satisfies: (1) $V_1 \subseteq V_2$, and $\forall v \in V_1$, $\theta_1(v)=\theta_2(v)$, (2) $E_1 \subseteq E_2$, and $\forall(u, v) \in E_1$, $\theta_1(u, v)=\theta_2(u, v)$. G_1 is an induced subgraph of G_2, if G_1 further satisfies $\forall u, v \in V_1$, $(u, v) \in E_1 \leftrightarrow (u, v) \in E_2$. G_2 can be denoted as a *supergraph* of G_1. A graph $G_1(V_1, E_1, L_{V1}, L_{E1}, \theta_1)$ is *isomorphic* to another graph $G_2(V_2, E_2, L_{V_2}, L_{E_2}, \theta_2)$, if and only if a bijection $f : V_1 \rightarrow V_2$ exists such that: (1) $\forall u \in V_1$, $\theta_1(u)=\theta_2(f(u))$, (2) $\forall(u, v) \in E_1 \leftrightarrow (f(u), f(v)) \in E_2$, (3) $\forall(u, v) \in E_1$, $\theta_1(u, v)=\theta_2(u, v)$. The bijection f is an isomorphism between G_1 and G_2. If there exists a subgraph $G_3 \subseteq G_2$ that a graph G_1 is isomorphic to G_3, we say that G_1 is subgraph isomorphic to G_2.

There are many ways to store graph information. The existing works aim to introduce more efficient strategies on designing effective representations on this issue. In the next several subsections we will briefly introduce these approaches.

6.4.2 Graph Representation

The common data structures used to store the graph information are adjacency matrix and adjacency list. For the adjacency matrix, the rows and columns denotes the vertices, and the intersection of row i and column j represents the edge between the vertices v_i and v_j. The value at the intersection $\langle i, j \rangle$ represents the number of edges between the vertices v_i and v_j. One main issue for adjacency matrix, is that it is difficult to detect the graph isomorphism by utilizing the matrix data structure. The reason is that there could be many possible adjacency matrices for the same graph by using different traversing strategies [52]. To address this problem, it is essential to guarantee that the same graphs (that may have different format) should be represented by the identical representation. Many studies explored this issue by introducing effective labeling strategies. To facilitate detecting the graph isomorphism, many researchers proposed to represent the graphs by using a unified coding strategy, i.e., canonical labeling [45]. The technique guarantees that if some graphs are isomorphic, their canonical labeling representation will be the same [32]. To fulfill this purpose, a reasonable approach is that we can utilize the common data structure, i.e., adjacency matrix, with some modification by taking into account the lexicographical ordering. Further optimization on compressing the canonical representation has been introduced [45]. For the remaining part of this section, we will introduce several canonical labeling strategies.

Depth First Search (DFS) Code: Yan et al. introduced the DFS code as a canonical labeling strategy [55]. Each edge in the graph encoded by DFS code is represented as $(v_i, v_j, l_{v_i}, l_{v_j}, l_e)$, where v_i and v_j are the vertices, l_{v_i} and l_{v_j} are the labels for l_{vi} and lvj, l_e is the label for the edge linking l_{v_i} and l_{v_j}. The basic method of DFS coding is that while traversing the graph according to the depth first search order, the vertex is labeled by a unique identifier sequentially. There could be several kinds of DFS codes and the existing works always aim at introducing an identical effective labeling strategy [55].

Canonical Adjacency Matrix: Inokuchi et al. proposed the canonical adjacency matrix (*CAM*) as the unique representation of the graph [26]. *CAM* can be obtained by encoding an adjacency matrix *AM* of a graph, through concatenating the lower triangular entries of *AM* which also takes into account of the diagonal. Because there can be many possible representations of the adjacency matrices, *CAM* is the one with the maximal or minimal encoding, which uniquely represents the graph information.

6.4.3 Candidate Generation

Because almost all the algorithms follow the candidate generate-and-test strategy, in the subsection we introduce how to deal with the candidate generation, which is an important step during the mining process. The challenge is that how to generate the candidates without redundancy and none of them is missing.

6.4.3.1 Join Operation

Borrowing from the idea of SPADE [62], Kuramochi et al. [32] introduced the join operation to generate the candidate subgraphs. The key idea in [32] is that two frequent k^2-subgraphs which have the same $(k-1)$-subgraph are joined, to generate a $(k+1)$-subgraph candidate. The main challenge, however, is that there could be many candidates produced. The reason is that a k-subgraph may have k different $(k-1)$-subgraphs. Kuramochi et al. tackled this issue by making a constraint that the $(k-1)$-subgraphs should be the two $(k-1)$-subgraphs which have the smallest and the second smallest canonical labels. This constraint largely reduces the candidates necessary to be generated and therefore, improves the whole performance.

---

[2]Here k could be the number of vertices or edges.

6.4.4 Frequent Subgraph Mining Algorithms

Because of the importance of frequent subgraph mining, there has been many algorithms proposed to tackle the issue. Similar to that of itemeset, sequence, and tree mining, the candidate generation and test are the main issues during the mining process. It is well known that the problem of detecting graph isomorhpism is NP-complete and therefore, the existing works aim to introduce efficient heuristic techniques to reduce the complexity of the problem. Detail surveys of the frequent graph mining in the literature can be found in [52, 22].

6.4.4.1 Apriori-based Graph Mining (AGM) Algorithm

AGM [26] is recognized as the first algorithm proposed to tackle the issue of frequent graph mining. The basic idea of AGM is that it applies the classic Apriori property to facilitate the mining process. To represent the graph, AGM uses the adjacency matrix. The join operation, therefore, can be executed straightforward by using basic matrix computation. According to the paper [26], the performance of AGM on real data (i.e., chemical data) has confirmed to be more efficient than the state-of-the-art approach. Moreover, the resultant patterns include those useful and undiscovered ones by previous work. In a later paper [27], the same authors explored the issue of frequent graph mining on more categories of graph data (i.e., directed v.s. undirected, labeled v.s. unlabeled, loop graph).

6.4.4.2 Frequent Subgraph Mining Algorithm (FSG)

The basic idea of FSG [32] also follows the Apriori rule. The main distinct features of FSG are that [32]: (1) a sparse graph representation which minimizes both storage and computation has been introduced; (2) when we generate the candidate, FSG adds one edge at a time to the discovered frequent subgraph; (3) the proposed algorithms are simple and the graph isomorphism detection are efficient for small graphs; and (4) some optimization are introduced to scale the algorithm on large graph data. A main issue, however, is that FSG does not perform well on very large data, as illustrated in the experimental evaluation.

6.4.4.3 Path Mining (PM) Algorithm

Gudes et al. introduced the path mining algorithm [21], which uses edge-disjoint paths as the expansion units to generate the candidate subgraphs. The key idea of *PM* is to decrease the number of the candidate patterns as early as possible. Furthermore, it minimizes the number of expensive support computations. There are several steps executed in the *PM* algorithm. It first extracts all the frequent paths, and then it discovers all the subgraphs which have two paths. Finally it joins the frequent subgraphs with (k-1)-paths which have the same (k-2)-paths, to obtain the candidate k-paths subgraph. Similar to other pattern mining algorithm papers, Gudes et al. stated that the support computation is the most time consuming step.

6.4.4.4 Graph-based Substructure Pattern Mining Algorithm (gSpan)

Yan et al. introduced the gSpan algorithm [55], which utilizes the *DFS* code to uniquely represent the graph. *gSpan* applies the depth first search strategy to traverse all the candidate subgraphs in the whole lattice which constructs a DFS code tree, whose nodes are the corresponding DFS code. The algorithm traverses the DFS code tree and all the subgraphs that have not minimal DFS codes are removed. By this way, it can avoid to generate the redundant candidates. Moreover, *gSpan* only saves the transaction lists for the discovered patterns and scans these lists to detect subgraph isomorphism. As shown in the paper [55], *gSpan* is efficient on both time and space cost compared with the state-of-the-art techniques.

6.4.4.5 Fast Frequent Subgraph Mining Algorithm (FFSM)

Huan et al. introduced another efficient algorithm, FFSM [25], to tackle the issue of frequent graph mining. The basic idea of FFSM is that it utilizes a vertical search scheme within an algebraic graph framework to reduce the number of redundant candidates tested. There are several distinct features in FFSM: (1) a novel graph canonical form and two efficient candidate generation operations, i.e., join and extension; (2) an algebraic graph framework (suboptimal CAM tree) to guarantee that all frequent subgraphs are enumerated unambiguously; and (3) avoid to test subgraph isomorphism by maintaining an embedding set for each frequent subgraph. The experimental evaluation demonstrates that FFSM outperformed gSpan on several chemical data sets [25].

It is very difficult to explore the advantages and disadvantages of the various frequent graph mining algorithms, because they are incomparable from many aspects. However, there are still some works towards this purpose, i.e., [53].

6.5 Chapter Summary

In this chapter, we have discussed the issues related to frequent pattern mining (i.e., itemset, sequence, tree, graph). These problems are fundamental issues in the data mining literature and are the basis of many practical applications. Some strategies have been commonly utilized by different algorithms (i.e., Apriori rule). Although so many works have studied the problem of frequent pattern mining, there are still many challenges existing. To name a few: (1) How to deal with huge data?; (2) How to mine exact patterns from stream data?; and (3) How to judge the usefulness and effectiveness of those discovered patterns. It seems that there is a still long way to reach the original goal of the data mining research.

References

[1] R. Agrawal, T. Imielinski and A. N. Swami. Mining association rules between sets of items in large databases. *In: Proceedings of the ACM SIGMOD International Conference on Management of Data*, pp. 207–216, 1993.

[2] R. Agrawal, H. Mannila, R. Srikant, H. Toivonen and A. I. Verkamo. Fast discovery of association rules. *Advances in Knowledge Discovery and Data Mining*, pp. 307–328, American Association for Artificial Intelligence, 1996.

[3] R. Agrawal and R. Srikant. Fast algorithms for mining association rules. *In:* Proceedings of *International Conference on Very Large Data Bases*, pp. 487–499, 1994.

[4] R. Agrawal and R. Srikant. Fast algorithms for mining association rules in large databases. *In: Proceedings of the 20th International Conference on Very Large Data Bases*, pp. 487–499, 1994.

[5] R. Agrawal and R. Srikant. Mining sequential patterns. *In: Proceedings of International Conference on Data Engineering*, pp. 3–14, 1995.

[6] T. Asai, K. Abe, S. Kawasoe, H. Arimura, H. Sakamoto and S. Arikawa. Efficient substructure discovery from large semi-structured data. *In: Proceedings of the Second SIAM International Conference on Data Mining*, 2002.

[7] J. Ayres, J. Gehrke, T. Yiu and J. Flannick. Sequential pattern mining using a bitmap representation. *In: Proceedings of ACM SIGKDD International Conference on Knowledge Discovery and Data Mining*, pp. 429–435, 2002.

[8] T. L. Bailey and C. Elkan. Fitting a mixture model by expectation maximization to discover motifs in biopolymers. *In: Proceedings of International Conference on Intelligent Systems for Molecular Biology*, pp. 28–36, 1994.

[9] R. J. Bayardo. Efficiently mining long patterns from databases. *In: Proceedings of the ACM SIGMOD International Conference on Management of Data*, pp. 85–93, 1998.

[10] S. Brin, R. Motwani and C. Silverstein. Beyond market baskets: generalizing association rules to correlations. *In: Proceedings of the ACM SIGMOD International Conference on Management of Data*, pp. 265–276, 1997.

[11] Y. Chi, R. R. Muntz, S. Nijssen and J. N. Kok. Frequent subtree mining—An overview. *Fundamenta Informaticae*, 66: 161–198, November 2004.

[12] Y. Chi, Y. Xia, Y. Yang and R. R. Muntz. Mining closed and maximal frequent subtrees from databases of labeled rooted trees. *IEEE Trans. on Knowledge and Data Engineering*, 17: 190– 202, 2005.

[13] Y. Chi, Y. Yang and R. R. Muntz. Indexing and mining free trees. *In: Proceedings of the Third IEEE International Conference on Data Mining*, 2003.

[14] Y. Chi, Y. Yang and R. R. Muntz. Hybridtreeminer: An efficient algorithm for mining frequent rooted trees and free trees using canonical forms. *In: Proceedings of the 16th International Conference on Scientific and Statistical Database Management*, 2004.

[15] D. J. Cook and L. B. Holder. *Mining Graph Data*. John Wiley & Sons, 2006.

[16] G. Dong and J. Li. Efficient mining of emerging patterns: discovering trends and differences. *In: Proceedings of ACM SIGKDD International Conference on Knowledge Discovery and Data Mining*, pp. 43–52, 1999.

[17] E. Eskin and P. Pevzner. Finding composite regulatory patterns in dna sequences. *In: Proceedings of the International Conference on Intelligent Systems for Molecular Biology*, pp. 354–363, 2002.

[18] U. M. Fayyad, G. Piatetsky-Shapiro, P. Smyth and R. Uthurusamy. *Advances in Knowledge Discovery and Data Mining*. AAAI/MIT Press, 1996.

[19] M. N. Garofalakis, R. Rastogi and K. Shim. Spirit: Sequential pattern mining with regular expression constraints. *In: Proceedings of International Conference on Very Large Data Bases*, pp. 223–234, 1999.

[20] B. Goethals. Survey on frequent pattern mining. Technical report, 2002.

[21] E. Gudes, S. E. Shimony and N. Vanetik. Discovering frequent graph patterns using disjoint paths. *IEEE Trans. on Knowl. and Data Eng.*, 18: 1441–1456, November 2006.

[22] J. Han, H. Cheng, D. Xin and X. Yan. Frequent pattern mining: current status and future directions. *Data Min. Knowl. Discov.*, 15: 55–86, August 2007.

[23] J. Han, G. Dong and Y. Yin. Efficient mining of partial periodic patterns in time series database. *In: Proceedings of International Conference on Data Engineering*, pp. 106–115, 1999.

[24] J. Han, J. Pei and Y. Yin. Mining frequent patterns without candidate generation. *In: Proceedings of the 2000 ACM SIGMOD International Conference on Management of Data*, pp. 1–12, 2000.

[25] J. Huan, W. Wang and J. Prins. Efficient mining of frequent subgraphs in the presence of isomorphism. *In: Proceedings of the Third IEEE International Conference on Data Mining*, 2003.

[26] A. Inokuchi, T. Washio and H. Motoda. An apriori-based algorithm for mining frequent substructures from graph data. *In: Proceedings of the 4th European Conference on Principles of Data Mining and Knowledge Discovery*, PKDD'00, pp. 13–23, 2000.

[27] A. Inokuchi, T. Washio and H. Motoda. Complete mining of frequent patterns from graphs: *Mining graph data. Machine Learning*, 50: 321–354, March 2003.

[28] C. Jiang, F. Coenen, R. Sanderson and M. Zito. A survey of frequent subgraph mining algorithms. *The Knowledge Engineering Review*, 00: 1–31, 2004.

[29] M. Kamber, J. Han and J. Chiang. Metarule-guided mining of multi-dimensional association rules using data cubes. *In: Proceedings of ACM SIGKDD International Conference on Knowledge Discovery and Data Mining*, pp. 207–210, 1997.

[30] M. Klemettinen, H. Mannila, P. Ronkainen, H. Toivonen and A. I. Verkamo. Finding interesting rules from large sets of discovered association rules. *In: Proceedings of ACM Conference on Information and Knowledge Management*, pp. 401–407, 1994.

[31] H. C. Kum, J. Pei, W. Wang and D. Duncan. Approxmap: Approximate mining of consensus sequential patterns. *In: Proceedings of SIAM International Conference on Data Mining*, pp. 311–315, 2003.

[32] M. Kuramochi and G. Karypis. Frequent subgraph discovery. *In: Proceedings of the 2001 IEEE International Conference on Data Mining*, pp. 313–320, 2001.

[33] B. Lent, A. Swami and J. Widom. Clustering association rules. *In: Proceedings of International Conference on Data Engineering*, pp. 220–231, 1997.

[34] C. Luo and S. Chung. Efficient mining of maximal sequential patterns using multiple samples. *In: Proceedings of SIAM International Conference on Data Mining*, pp. 64–72, 2005.

[35] N. Mabroukeh and C. Ezeife. A taxonomy of sequential pattern mining algorithms. *ACM Computing Surveys*, 2010.

[36] N. R. Mabroukeh and C. I. Ezeife. A taxonomy of sequential pattern mining algorithms. *ACM Computing Survey*, 43(1): 3:1–3:41, Dec 2010.

[37] H. Mannila, H. Toivonen and A. I. Verkamo. Discovering frequent episodes in sequences. *In: Proceedings of ACM SIGKDD International Conference on Knowledge Discovery and Data Mining*, pp. 210–215, 1995.

[38] H. Mannila, H. Toivonen and A. I. Verkamo. Discovery of frequent episodes in event sequences. *Data Mining and Knowledge Discovery*, 1(3): 259–289, 1997.

[39] H. Mannila, H. Toivonen and I. Verkamo. 1994. Efficient algorithms for discovering association rules. *In: Proceedings of the AAAI Workshop on Knowledge Discovery in Databases*, pp. 181–192, 1994.

[40] S. Nijssen and J. N. Kok. A quickstart in frequent structure mining can make a difference. *In: Proceedings of the Tenth ACM SIGKDD International Conference on Knowledge Discovery and Data Mining*, pp. 647–652, 2004.

[41] J. S. Park, M.-S. Chen and P. S. Yu. An effective hash-based algorithm for mining association rules. *In: Proceedings of the ACM SIGMOD International Conference on Management of Data*, pp. 175–186, 1995.

[42] S. Parthasarathy, M. J. Zaki, M. Ogihara and S. Dwarkadas. Incremental and interactive sequence mining. *In: Proceedings of ACM Conference on Information and Knowledge Management*, pp. 251–258, 1999.

[43] J. Pei, J. Han, B. Mortazavi-Asl and H. Pinto. Prefixspan:mining sequential patterns efficiently by prefix-projected pattern growth. *In: Proceedings of International Conference on Data Engineering*, pp. 215–224, 2001.

[44] J. Pei, J. Han, B. Mortazavi-Asl, J.Wang, H. Pinto, Q. Chen, U. Dayal and M. Hsu. Mining sequential patterns by pattern-growth: The prefixspan approach. *IEEE Transactions on Knowledge and Data Engineering*, 16(11): 1424–1440, November 2004.

[45] R. Read and D. Corneil. The graph isomorphism disease. *Journal of Graph Theory*, 1: 339–363, July 1977.

[46] C. Silverstein, S. Brin, R. Motwani and J. Ullman. Scalable techniques for mining causal structures. *Data Mining and Knowledge Discovery*, 4(2-3): 163–192, 2000.

[47] R. Srikant and R. Agrawal. Mining sequential patterns: Generalizations and performance improvements. *In: Proceedings of International Conference on Extending Database Technology*, pp. 3–17, 1996.

[48] W. G. Teng, M. Chen and P. Yu. A regression-based temporal pattern mining scheme for data streams. *In: Proceedings of International Conference on Very Large Data Bases*, pp. 93–104, 2003.

[49] P. Tzvetkov, X. Yan and J. Han. Tsp: Mining top-k closed sequential patterns. *In: Proceedings of IEEE International Conference on Data Mining*, pp. 347–358, 2003.

[50] K. Wang and H. Liu. Discovering typical structures of documents: a road map approach. *In: Proceedings of the 21st Annual International ACM SIGIR Conference on Research and Development in Information Retrieval*, pp. 146–154, 1998.

[51] W. Wang and J. Yang. *Mining Sequential Patterns from Large Data Sets*, Vol. 28. Series: The Kluwer International Series on Advances in Database Systems, 2005.

[52] T. Washio and H. Motoda. State of the art of graph-based data mining. *SIGKDD Explor. Newsl.*, 5: 59–68, July 2003.

[53] M. Wörlein, T. Meinl, I. Fischer and M. Philippsen. A quantitative comparison of the subgraph miners mofa, gspan, ffsm, and gaston. *In: Proceedings of the 9th European Conference on Principles and Practice of Knowledge Discovery in Databases*, pp. 392– 403, 2005.

[54] Y. Xiao, J.-F. Yao, Z. Li and M. H. Dunham. Efficient data mining for maximal frequent subtrees. *In: Proceedings of the Third IEEE International Conference on Data Mining*, 2003.

[55] X. Yan and J. Han. gspan: Graph-based substructure pattern mining. *In: Proceedings of IEEE International Conference on Data Mining*, pp. 721–724, 2001.

[56] X. Yan, J. Han and R. Afshar. Clospan: mining closed sequential patterns in large datasets. *In: Proceedings of SIAM International Conference on Data Mining*, pp. 166– 177, 2003.

[57]   Z. Yang and M. Kitsuregawa. Lapin-spam: An improved algorithm for mining sequential pattern. *In: Int'l Special Workshop on Databases For Next Generation Researchers (SWOD)*, pp. 8–11, 2005.

[58]   Z. Yang, Y. Wang and M. Kitsuregawa. Effective sequential pattern mining algorithms for dense database. *In: National Data Engineering WorkShop (DEWS)*, 2006.

[59]   Z. Yang, Y. Wang and M. Kitsuregawa. An effective system for mining web log. *In: The 8th Asia-Pacific Web Conference (APWeb)*, pp. 40–52, 2006.

[60]   Z. Yang, Y. Wang and M. Kitsuregawa. Lapin: Effective sequential pattern mining algorithms by last position induction for dense databases. *In: Int'l Conference on Database Systems for Advanced Applications (DASFAA)*, pp. 1020–1023, 2007.

[61]   M. J. Zaki. Scalable algorithms for association mining. *IEEE Transaction on Knowledge and Data Engineering*, 12(3): 372–390, 2000.

[62]   M. J. Zaki. Spade: An efficient algorithm for mining frequent sequences. *Machine Learning Journal*, 42: 31–60, 2001.

[63]   M. J. Zaki. 2002. Efficiently mining frequent trees in a forest. *In: Proceedings of the eighth ACM SIGKDD international conference on Knowledge discovery and data mining*, pp. 71–80.

[64]   M. J. Zaki, S. Parthasarathy, M. Ogihara and W. Li. 1997. New algorithms for fast discovery of association rules. In *Proceedings of ACM SIGKDD International Conference on Knowledge Discovery and Data Mining*, pp. 283–286.

Part II
Advanced Data Mining

CHAPTER 7

Advanced Clustering Analysis

7.1 Introduction

Clustering is a useful approach in data mining processes for identifying patterns and revealing underlying knowledge from large data collections. The application areas of clustering include image segmentation, information retrieval, and document classification, associate rule mining, web usage tracking and transaction analysis. Generally, clustering is defined as the process of partitioning unlabeled data set into meaningful groups (clusters) so that intra-group similarities are maximized and inter-group similarities are minimized at the same time. In essence, clustering involves the following unsupervised learning process, which can be written as: Define an encoder function c(x) to map each data object xi into a particular group $Gk(c(x) = k) \Rightarrow x \in Gk$, $k = 1, 2, 3, ..., k$, so that a cluster criterion $Q(c) = \sum_k = 1^K \sum_c (x_i) = k$, $c(x_j = k)dist(x_i, x_j)$ is minimized.

As we know, this is a classical combinatorial optimization problem and solving it is exactly NP-hard, even with just two clusters [13]. According to computation complexity theory [36], no complete algorithm can get the overall optimal solutions in a polynomial time, unless P = NP. Iterative refinement method, a popular approximate algorithm, is widely adopted by various unsupervised learning algorithms. A general iterative refinement clustering process can be summarized as Algorithm 7.1 [6].

Algorithm 7.1: General iterative refinement clustering

Initialization: Initialize the parameters of the current cluster model.

Refinement: Repeat until the cluster model converges.
  (1) Generate the cluster membership assignments for all data objects, based on the current model;
  (2) Refine the model parameters based on the current cluster membership assignments.

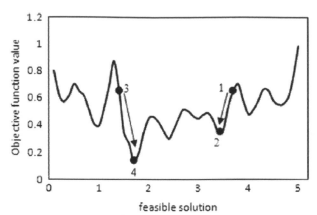

Figure 7.1.1: An example of iterative refinement clustering algorithm

The intuitionists denotation of iterative refinement clustering algorithm is shown in Fig. 7.1.1. The horizontal axis denotes feasible solutions of clustering problem and the vertical axis is the corresponding objective function values of feasible solutions.

In this chapter, the feasible solution is the results of encode function (or the clustering results) and the objective function value is the values of cluster criterion $Q(c) = \Sigma_k = 1^K \Sigma_c (x_i) = k, c(x_j = k)dist(x_i, x_j)$. Without loss of generality, we assume that point 3 is selected as the initialization of an iterative refinement clustering algorithm, and by repeating step (1) and (2), the algorithm will converge to point 4, one of the feasible solutions with suboptimal objective function value. An iterative refinement clustering algorithm is a popularly used clustering approach based on heuristic search, which is to minimize the cluster criterion in a designated region (i.e., guarding the heuristic search with the specified initial values). In this scenario, the obtained clustering result is dependent on the initialization and the minimized cluster criterion only reflects the sub-optimal solution with this running of heuristic search. In other words, the nature of heuristic search process makes the iterative refinement clustering algorithms heavily sensitive to the initialization settings, thus not guaranteeing the higher quality clustering results with randomly chosen initializations [5]. Therefore, how to deal with the sensitivity problem of initialization in iterative refinement clustering algorithm is becoming an active and well concerned. That the initialization model must be correct is an important underlying assumption for iterative refinement clustering algorithm. It can determine the clustering solution [6], that is, different initialization models will produce different clustering results (or different local minimum points as shown in Fig. 7.1.1). Since the problem of obtaining a globally optimal initial state has been shown to be NP-hard [15], the study on the initialization methods towards a sub-optimal clustering result hence is more practical, and is of great value for the clustering research. Recently, initialization methods have been categorized into three major families: random sampling methods, distance optimization methods and density estimations [17]. Forgy adopted uniformly random input objects as the seed clusters [14], and MacQueen gave an equivalent way with selecting the first K input objects as the seed clusters [29]. In the FASTCLUS, a K-means variance implemented in SAS [37], the simple cluster seeking (SCS) initialization method is adopted [21]. Katsavounidis et al. proposed a method that utilizes the sorted pairwise distances for initialization [22]. Kaufman and Rousseeuw introduced a method that estimates the density through pairwise distance comparison, and initializes the seed clusters using the input objects from areas with high local density [23]. In Ref. [18], a method which combines local density approximation and random initialization is proposed. Belal et al. find a set of medians extracted from a dimension with maximum and then use the medians as the initialization of K-means [4]. Niu et al. give a novel algorithm called PR (Pointer Ring), which initializes cluster centers based on pointer ring by partition traditional hyper-rectangular units further to hyper-triangle subspaces [33]. The initialization steps of K-means++ algorithm can be described as: choosing an initial center m_1 uniformly at random from data set; and then selecting the next center $m_i = x_0$ from data set with a probability, where $dist(x, m)$ denote the shortest distance from a data object x to the closest center m; iterative until find K centers [2]. The main steps of initialization centers of K-means by kd-tree are: first, the density of a data at various locations are estimated by using kd-tree; and then use a modification of Katsavounidis algorithm, which incorporates this density information, to choose K seeds for K-means algorithm [39]. And recently, Lu et al. treated the clustering problem as a weighted clustering problem so as to find a better initial cluster center based on the hierarchical approach [28].

7.2 Space Smoothing Search Methods in Heuristic Clustering

The goal of modified initialization methods, is to reduce the influence of sub-optimal solutions (the local minimum points) bestrewed in the whole search space, as shown in Fig. 7.1.1. Although iterative refinement clustering algorithms with these modified initialization methods have some merits in improving the quality of cluster results, they also have high probability to be attracted by local minimum points. Local search method is the essence of iterative refinement clustering algorithms. Lots of the local minimum points make a local search problem hard and sensitive to the initialization. Those proposed modified initialization methods are only focused on how to select an initialization which can improve the quality of iterative refinement clustering algorithm, but the search space embedded lots of local minimum points is ignored. Smoothing search space method reconstructs the search space by filling local minimum points, to reduce the influence of local minimum points. In this paper, we first design two smoothing operators to reconstruct the search space by filling the minimum traps (points) based on the relationship between distance metric and cluster criterion. Each smoothing operator has a parameter, smoothing factor, to control the number of minimum traps. And then, we give a topCdown clustering algorithm with smoothing search space (TDCS3) to reduce the influence of initialization. The main steps of TDCS3 are to: (1) dynamically reconstruct a series of smoothed search space as a hierarchical structure: the most smoothed search space at the top, and the original search space at the bottom, other smoothed search spaces are distributed between them, by filling the local minimum points; (2) at the top level of the hierarchical structure, an existing iterative refinement clustering algorithm is run with random initialization to generate the cluster result; (3) from the second level to the bottom level of the hierarchical structure, the same clustering algorithm is run with the initialization derived from the cluster result on the previous level.

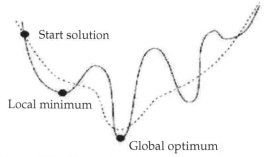

Figure 7.2.1: Illustration of smoothing search space

7.2.1 Smoothing Search Space and Smoothing Operator

7.2.1.1 Local Search and Smoothing Search Space

Local search method is the essence of iterative refinement clustering algorithms. During the mid-sixties, local search method was first proposed to cope with the overwhelming computational intractability of NP-hard combinatorial optimization problems. Give a minimization (or maximization) problem with objective function f and feasible region F, a typical local search algorithm requires that, with each solution $x_i \in R^d$, there is associated a predefined neighborhood $N(x_i) \subset R^d$. Given a current solution point $x_i \in R^d$, the set $N(x_i)$ is searched for a point x_i+1 with $f(x_i + 1) < f(-x_i)$ or $(f(x_i + 1) > f(x_i))$. If such a point exists, it becomes the new current solution point $(x_i \, x_i+1)$, and then the process is iterated. Otherwise, x_i is retained as a local optimum with respect to $N(x_i)$. Then a set of feasible solution points is generated, and each of them is locally improved within its neighborhood. Local search methods only check the neighborhood of current feasible solution x_i, so the search range has been dramatically reduced and the convergence speed has been accelerated. A major shortcoming of local search is that the algorithm has a tendency to get stuck at a locally optimum configuration, i.e., a local minima point, as the point 2 or 4 shown in Fig. 7.1. Different neighborhood structures result in difference terrain surface structures of the search space and produce different numbers of local minimum points. The effectiveness of a local search algorithm relies on the number of local minimum points in the search space [16], that is, local minimum points make a search problem hard. The smaller the number of local minimum points, the more effective a local search algorithm is. In order to reduce the influence of local minimum to local search algorithm, some local minimum traps must be filled. Gu and Huang [16] has called the method of filling minimum trap as the smoothing search space, and it is able to dynamically reconstruct the problem structure and smooth the rugged terrain surface of the search space. The smoothed search space could hide some local minimum points, therefore, improving the performance of the traditional local search algorithm. Figure 7.2.1 is the illustration of smoothing search space.

From Fig. 7.2.1, we can see that Many local minimum traps are filled after running a smoothing operator. The real line curve shows the original search space which has lots of minimum traps, and dashes shows the smoothed search space with fewer minimum traps. At the former discussing, we can find that lots of the local minimum points which are embedded in the search space make a local search problem hard and sensitive to the initialization. The essence of iterative refinement clustering algorithms is the local search method, thus they have the same real reason for initialization sensitivity problem. The main idea of smoothing search space is always common, but different application areas have different ways to smoothing the search space. In clustering area, clustering is defined as the process of partitioning unlabeled data objects into meaningful groups (clusters) so that the value of cluster criterion $Q(c)$ is minimized. Minimizing $Q(c)$ value means that the intra-similarities of all clusters are maximized or the distances of each data object to its cluster center is minimized. So the cluster criterion $Q(c)$ has a close relationship with the similarity or distance between data objects.

7.2.1.2 Smoothing Operator

In this section, we designed two smoothing operators based on the relationship between $Q(c)$ and distance measure, to fill the minimum traps embedded in the rugged surface of search space. Let $D = x_1, x_2, \dots x_N, x_i \in R^d$ be a set of data objects that needs to be clustered. And note dist:$R^d \times R^d \rightarrow R_+$ be a given distance function between any two data objects in R^d. Dist is a distance matrix which contains the distances between all data objects of D, and $Dist(i,j)$ denotes the distance between data object x_i and x_j, $Dist(x_i,x_j)$. In this chapter, two smoothing operators will be described as follows:

(1) Displacement smoothing operator
     Based on average distance of distance matrix Dist, we design the displacement smoothing operator as below.

Definition 7.1. Given a data set $D = x_1, x_2, \dots x_N$, and its distance matrix Dist, the average distance of Dist is defined as: $\overline{Dist} = \dfrac{1}{N(N-1)} \Sigma_i = 1^N \Sigma_j = 1^N Dist(i, j)$.

Definition 7.2. Given a smoothing factor $\alpha \geq \alpha_a rg$, the displacement smoothing operator reconstructs the smoothed search space according to:

$$Dist_\alpha(i, j) = \begin{cases} \overline{Dist} + (Dist(i, j) - \overline{Dist})^\alpha & \text{if} \quad Dist(i, j) \geq \overline{Dist} \\ \overline{Dist} - (\overline{Dist} - Dist(i,j))^\alpha & \text{if} \quad Dist(i, j) < \overline{Dist} \end{cases}$$

According to Definitions 7.1 and 7.2, a series of smoothed search spaces with different numbers of minimum traps will be reconstructed during $\alpha \rightarrow \alpha_a rg$. A smoothed search space generated from a large exhibits a smoother terrain surface, and a search space generated from a smaller exhibits a more rugged terrain surface. The search space will return to the original search space when $\alpha = \alpha_a rg$. Lets note the smoothed search space according to the largest serves as the top search space and the original search space as the bottom search space, as shown in Fig. 7.2.2.

Algorithm 7.2 describes the details of the reconstruction process for smoothing the search spaces. In the first step, we calculate the average distance of and during the second step, a distance transformation is run to change each distance $Dist(i, j) \in Dist$ with average distance $\overline{Dist}$ and the difference between $dist(x_i, x_j)$ and $\overline{Dist}$. The main time cost of displacement smoothing operator is the process of the distance transformation. For a distance $Dist(i, j) \in Dist$, the time cost of distance transformation is . For all the distances belong to , the total time consume is $O(N^2)$.

In this section, we set $\alpha_o rg = 1$, then there are two extreme cases of the series of the clustering instances, which are based on the distance. These are:

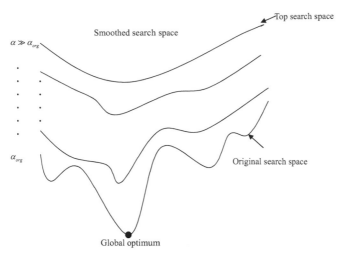

Figure 7.2.2: Illustration of a series smoothed search space with different terrain surfaces, which are generated by displacement methods with different smoothing factors

---

Algorithm 7.2: Displacement Smoothing Operator

Input: distance matrix $Dist$, smoothing factor α

Output: smoothed search space S^α

  (1) Calculate the average distance $\overline{Dist}$ of $Dist$;

  (2) For any $Dist(i, j) \in Dist$

      If $Dist(i, j) < \overline{Dist}$ then

$$Dist_\alpha(i, j) = \overline{Dist} - (\overline{Dist} - Dist(i, j))^\alpha;$$

      Else

$$Dist_\alpha(i, j) = \overline{Dist} + (Dist(i, j) - \overline{Dist})^\alpha;$$

      End if;

    End for;

  (3) $S^\alpha \leftarrow Dist_\alpha$ and return.

- if $\alpha \neq \alpha_0 rg$, then $Dist_\alpha \to \overline{Dist}$ this is the trivial case;
- if $\alpha = \alpha_0 rg$, then $Dist_\alpha = \overline{Dist}$, which is the original problem.

(2) Kernel Smoothing Operator

The main idea of the displacement smoothing operator is the linear transformation of distance based on $\overline{Dist}$ and the exponential of the difference between $dist(x_i, x_j)$ and $\overline{Dist}$. This smoothing operator fits well to linear problem, but is weak with the non-linear problem. So another smoothing operator which could be extended to non-linear situation is designed. This smoothing operator, named kernel smoothing operator, is based on the smoothing kernel.

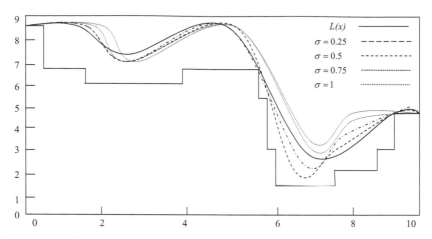

Figure 7.2.3: Illustration of a smoothing space after running a kernel smoothing operator on piece-wise function

DEFINITION 7.3 [3]. Given a real value function $f: R^n \rightarrow R$ and a smoothing kernel $g : R \rightarrow R$, which is a continuous, bounded, nonnegative, and symmetric function whose integral is one, the g-transform of is defined as $\langle f \rangle_g(x) = \int (R^n) f(y) g(\mid\mid y - x \mid\mid) dy$. The Gaussian kernel function $g(z) = exp(-z^2/(2\sigma^2))$ is the most widely used kernel. Figure 4 gives an example of applying a smoothing transformation to the piecewise constant function and we estimate the transformed function. $\langle L \rangle_g(x) = \int (R^n) f(y) g(\mid\mid y - x \mid\mid) dy$.

From Fig. 7.2.3, we can see the traps of minimum point has been smoothed by the Gaussian kernel function with different σ, the smoothing factor.

In Fig. 7.2.3, the real line curve illustrates the original search space, and all the dashed are the smoothed search space, by running a kernel smoothing operator with different smoothing factor σ on the original search space. In this chapter, we use the kernel smoothing method to smooth the distance function $dist(x_i, x_j)$ and reduce the influence of lots of minimum value embedded in the search space. We assume that there is no missing value in data set D, that is, distance function dist is a continuous function. Let $L(x) = dist(x_i, x_j) = \Sigma_i = 1^d \mid\mid x_i - x_j \mid\mid^2$, the smoothing method for clustering is defined as: $dist(x_i, x_j)_g = dist(x_i, x_j) * exp(-dist(x_i, x_j)^2/(2\sigma^2))$. The main steps of kernel smoothing operator are shown in Algorithm 3. For any pairwise data objects belong to data set D, a Gaussian kernel influence adds to distance function $dist(x_i, x_j)$ to smooth the surface of search space. Once transformation on a pairwise data object x_i, x_j needs O(1) time, so for the transformation of N^2 pairwise data objects $O(N^2)$ times is need at least.

Algorithm 7.3: Kernel Smoothing Operator

Input: data set D, smoothing factor σ

Output: smoothed search space S^α

(1) $S^\alpha = zeros(N, N)$;

(2) for any pairwise data objects, $x_i, x_j \in D$

$\qquad S^\alpha(i, j) = dist(x_i, x_j) * \exp(-dist(x_i, x_j)^2/2\sigma^2)$

(3) return S^α

7.2.2 Clustering Algorithm based on Smoothed Search Space

Based on the smoothing operator and the smoothing factor, a series of smoothed search space with different number of minimum traps, are reconstructed as a hierarchical structure. Iterative refinement clustering algorithm can be run on each smoothed search space from the top search space down to the bottom search space. An algorithm framework is proposed to realize this process in this section. For simple description, we use symbol α to denote the smoothing factor in the rest of this paper.

7.2.2.1 Top-Down Clustering Algorithm based on Smoothing Search Space

We give a Top-Down Clustering algorithm based on the Smoothed Search Space (TDCS3) in this section. The main ideas of TDCS3 are:

(1) dynamically reconstruct the smoothed search space by running a smoothing operator;

(2) run an existing iterative refinement clustering algorithm on the current smoothed search space and generate the cluster results;

(3) based on the cluster results, a new initialization is generated and services to the next smoothed search space;

(4) repeat run (1)–(3) until back to the original search space.

Algorithm 7.4 is the main description of the framework of TDCS3. From the top search space to the original search space, any existing iterative refinement clustering algorithm, such as k-means [13], MeanShift [41] and so on, could be run on these smoothed search space. At the top search space, algorithm Λ is run with random initialization, and the correlate cluster results are generated. The initialization from the former cluster results will be regards as the initialization of Λ in the current search space and lead the search of algorithm Λ to converge to a better sub-optimal result. Iteratively run these steps until reach the original search space.

7.2.2.2 Benefits

TDCS3 focuses on reducing the influence of lots of minimum trap embedded in the search space of iterative refinement clustering algorithm. Comparing to traditional iterative refinement clustering algorithm, TDCS3 has the following benefits:

(1) Intelligent characteristic under TDCS3 framework, a series of different smoothed search space are reconstructed. All the smoothed search spaces are the different level topological structures of the original search space. The quality of cluster results on more smoothed search space is high for the number of minimum trap more less than the original search space. So the initialization from the cluster result on more smoothed search space can capture good structure of clusters. The initialization from the cluster results on the former search space can lead the search on current search space to a better minimum point.

(2) Flexible characteristic TDCS3 is only an algorithm framework. The smoothing operator and the iterative clustering algorithm, which will be run under TDCS3, are not fixed. According to different applications and demands, the smoothing operator could be redesigned and the iterative refinement algorithm will be selected.

(3) Adaptive characteristic TDCS3 inherits the merits of iterative refinement clustering algorithm and reconstructs their search space to reduce the probability of getting stuck into a worse sub-optimal minimum points.

Algorithm 7.4: TDCS3

Input: Data set D, Cluster number K and Smoothing factor α

Output: cluster results C

(1) Generate the top search space S^α with α;

(2) Run any iterative refinement clustering Λ on S^α with random initialization, and generate the cluster results C^α;

(3) while $\alpha \geq \alpha_{org}$

    (3.1) Generate the initialization $Init^\alpha$ from C^α;

    (3.2) $\alpha' \leftarrow \alpha - \lambda$ and generate new search space $S^{\alpha'}$ with α';

    (3.3) Run Λ on $S^{\alpha'}$ with $Init^\alpha$, and generate the cluster results $C^{\alpha'}$;

    (3.4) $\alpha \leftarrow \alpha' \ C^\alpha \leftarrow C^{\alpha'}$;

(4) $C \leftarrow C^\alpha$ and return.

7.3 Using Approximate Backbone for Initializations in Clustering

As described in Section 7.2, iterative refinement clustering exactly is NP-hard, even with just two clusters. In real application, K-centre clustering algorithm is a traditional iterative refinement clustering, so it inherits the advantages and drawbacks from iterative refinement clustering.

For a larger data set, researchers are seeking heuristic methods to solve this clustering problem. For example, the K-centre clustering algorithm is a popularly used clustering approach based on heuristic search by minimizing the sum of squared error locally and obtaining the local suboptimal clustering results. However, the heuristic search process makes *K*-centre clustering algorithms heavily sensitive to the initialization, and usually cannot guarantee the high quality clustering results with random initializations [5]. On the other hand, different local suboptimal clustering results do reflect the different likelihood of data instances gathering around various centres in a data set [43]. Due to the fact that there are 80% local suboptimal solutions are observed to distribute around the global optimal solutions [24, 30, 35], it is believed that finding the commonly overlapped intersections of various local suboptimal clustering results will facilitate locating the global optimal solutions. Moreover, for *K*-centre heuristic clustering, it is expected that choosing these intersection areas as the initial search space will result in better clustering results. Backbone analysis is becoming an active research topic in NP-hard problem recently. The backbone of a NP-hard problem is regarded as the core part of all global optimal solutions, which was first proposed in [26] for Travelling Salesman Problem (TSP), and has attracted much attention recently [42, 38]. An exact backbone, however, is generally hard to be obtained for many optimization problems in real applications. Instead, Approximate Backbone (AB), as indicated by the name—the approximate form of backbone, and defined as the intersection of different local suboptimal solutions of a dataset, is often used to investigate the characteristic of the dataset and expedite the convergence speed of heuristic algorithms [44, 8, 19]. In this paper, we intend to adopt the concept of AB to address the initialization problems suffering the heuristic clustering described above, and in particular, propose a Heuristic Clustering Approach Based on Approximate Backbone (HC_AB). The basic idea of HC_AB is that: we first identify the AB from a set of local suboptimal solutions derived from running *K*-centre clustering with different initialization settings; then, construct a new restricted search space based on the AB for heuristic search; eventually re-run the K-centre clustering algorithm by using this new search space which has the AB as the part of initialization, and generate a better clustering result that is deemed to best approximate the global optimal solution.

7.3.1 Definitions and Background of Approximate Backbone

In this section, we give several definitions to prepare the background of approximate backbone. Given a data set $D = x_1, x_2, ...x_n$ contains N data objects and each object $x_i \in R^d$ is defined over-dimensional feature space. Let $dist : R^d \times R^d \mapsto R_+$ be a given distance function between any two objects in R^d. A K-centre clustering algorithm takes D as an input and partitions the N data objects into K clusters such that the sum of squared error $\phi = \sum_{k=1}^{K} \sum_{x_i \in C_k} dist(x_i, v_k)$ is minimized, where C_k is a cluster and v_k is the centre of C_k. Since each cluster is represented by its centre, the K-centre clustering result can be represented as $V = v_1, v_2, ...v_K$. The local suboptimal clustering result is represented by a set of centres such that the corresponding ϕ value is minimized in a local area. The Global optimal clustering results are the collection of suboptimal clustering results with the smallest ϕ value. DEFINITION 1 (Backbone). Given the global optimal clustering results $Z^* = V^1*, V^2*, ...V^p*$ where $V^p* = v_1^p*, v_2^p*, ...v_k^p*$, $p = 1, ...P$. The backbone of this clustering problem is defined as the intersection of P global optimal clustering results. $backbone(V^1*, V^2*, ...V^p*) = V^1* \cap V^2* \cap ....\cap V^p*$.

Generally, the global optimal solution is hard to be obtained for a NP-hard problem in fact, resulting in difficulty in identifying the theoretically ideal backbone. However, in many research areas, researchers have observed an interesting fact that there are 80% local suboptimal solutions being distributed around the global optimal solutions and a big valley structure is seen. Motivated by this fact, we intuitively have an idea in mind on how to approximate the ideal backbone by making use of the local suboptimal solutions.

DEFINITION 7.4 (Approximate Backbone). Given the local suboptimal clustering result $Z = V^1, V^2, ..., V^M$, where $V^m = v_1^m, v_2^m ..., v_K^m$ $m = 1, ...,M$. The AB is defined as the intersection of M local suboptimal clustering results.

$$a\_bone(V^1, V^2, ..., V^M) = V^1 \cap V^2 \cap .. \cap V^M$$

As described above, our method aims to use the AB of local optimal solutions to form the part of initialization (i.e., the start points for heuristic search), thus constructing an appropriate AB for the heuristic search in K-centre clustering algorithm is a key issue. In other words, the quality of K-centre clustering results is greatly dependent on the characteristics of AB. To address the concern, here we propose two parameters to describe the characteristics of AB—Scale and Purity. The former one describes how many percentages of total local optimal solutions are included in the AB; whereas the later one denotes how many percentages of local suboptimal solutions included in AB are also existed in the theoretically ideal backbone as well. In particular, Approximate Backbone Scale (ABS) and Approximate Backbone Purity (ABP) are defined as follows.

DEFINITION 7.5 (Approximate Backbone Scale). Given an AB, $a_b one$ $(V^1, V^2, ..., V^M)$. Approximate Backbone Scale is defined as the proportion of the AB cardinality to the cluster number K.

$$ABS = \frac{|a\_bone(V^1, V^2, ..., V^M)|}{K}$$

DEFINITION 7.6 (Approximate Backbone Purity). Given an AB, $a_b one(V^1, V^2, ..., V^M)$, and a backbone $backbone(V^{1*}, ..., V^{P*})$ Approximate Backbone Purity is defined as the proportion of the cardinality of the intersection of the AB and the backbone to the AB cardinality.

$$ABP = \frac{(|a\_bone(V^1, V^2, ..., V^M) \cap backbone(V^{1*}, V^{2*}, ..., V^{P*})|)}{|a\_bone(V^1, V^2, ..., V^M)|}$$

As the AB is used for the selection of initialization, in order to achieve the best result of heuristic search, we expect to form an appropriate AB with both large ABS and ABP values, which indicates the fact that the most of whole local suboptimal solutions should be included in the initialization and the included local suboptimal solutions (centres) are closely scattered around the global optimal clustering result. In order to better understand ABS and ABP, we give an example to explain them. Given a data set D, which contains 500 objects and 10 clusters, each cluster is represented by a representative object. An assumed global optimal clustering result V^* is located in the first row in table 1 and three local suboptimal clustering results V^1, V^2, V^3, obtained by running K-centre clustering algorithm with three initializations, are also listed in Table 7.3.1. For the simplification, we assume that each centre is represented by an objects ID in D, and there is only one global optimal clustering result in D, that is $backbone(V^*) = V^*$.

For this example, we can obtain the AB: $a_b one(V^1, V^2, V^3) = 43, 78, 198, 240, 310, 366, 480$. Known from the definition of ABS, the value of ABS in this example is calculated as follow.

$$ABS = \frac{|a\_bone(V^1, V^2, ..., V^3)|}{K} = \frac{7}{10} = 0.7$$

Name	Centre set
V^*	22, 78, 109, 180, 230, 292, 310, 366, 412, 475
V^1	43, 78, 109, 198, 240, 262, 310, 366, 412, 480
V^2	43, 78, 128, 198, 240, 262, 310, 366, 412, 480
V^3	43, 78, 128 198, 240, 252, 310, 366, 432, 480

Figure 7.3.1: Clustering results of D

We observed that there are three commonly overlapped centres existed in AB and backbone, thus the value of *ABP* is,

$$ABP = \frac{|a\_bone(V^1,V^2,V^3) \cap V^*|}{|a\_bone(V^1,V^2,V^3)|} = \frac{3}{7} = 0.429$$

According to the Definition 7.5, the *AB* is derived from *M* local suboptimal solutions, so the characteristics of AB has a close relationship with *M*. In order to illustrate this relationship, we construct three data sets: RandomS1, RandomS2 and RandomS3, each of which contains 34 clusters. And each cluster has 100 data objects, among which 99 objects are generated by a Gaussian distribution function with different mean (μ) and standard deviation (σ) and the last one is the mean of the rest, which is deemed as the centre of the cluster. We run Vertex Substitution Heuristic (VSH) algorithm [7], a classical *K*-centre clustering algorithm, on these three data sets, and note the process as $VSH_R andomS1$, $VSH_R andomS2$ and $VSH_R andomS3$ respectively. VSH was executed for *M*=2:2:20 times on each data set, where *M*=2:2:20 means *M* changing from 2 to 20 with step 2. The relationships between *ABS*, *ABP* and *M* are shown in Fig. 7.3.2.

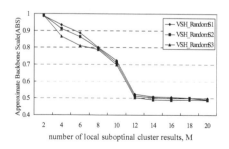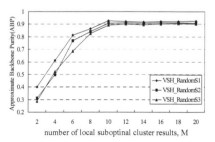

Figure 7.3.2: The ABS and ABP of AB

From Fig. 7.3.2, we can see that the trends of changes on *ABS* and *ABP* are along opposite directions with M—*ABS* is decreasing while M increasing, on the contrary, ABP increasing, and eventually, the changes of *ABS* and *ABP* become slight and a balanced state is reached. As indicated above, we hope to form a better *AB* with both higher *ABS* and *ABP* values to construct a good searching start for *K*-centre clustering algorithm. Due to the inconsistent changes of *ABS* and *ABP* with *M*, we have to choose a tradeoff between *ABS* and *ABP* to ensure the better clustering results. According to the experimental results, we find the fact that choosing a reasonable *M* can guarantee a better *AB*. The setting of *M* is an open question in various applications [19]. In this chapter, we experimentally set M value based on the size of data set and the cluster number.

7.3.2 Heuristic Clustering Algorithm based on Approximate Backbone

7.3.2.1 Reconstruct the New Searching Space

A new search space, *S* determined by the *AB*, is crucial to heuristic clustering algorithms. Various applications might have different new search space reconstruction methods. For example, in *TSP* [16], the new search space is reconstructed by including all the edges that are not occurred in *AB*. While the construction of *AB* leads to the formation of the fraction of the new search space, the number of centres within the *AB* is usually less than the number of predefined clusters in *K*-centre clustering. In that case, how to select the rest centres from the local suboptimal solutions is another important issue. Essentially, a cluster centre that was'nt included in the AB but was frequently occurred in local suboptimal clustering results is more likely to be selected to represent a real cluster. Although we can refine the new search space with all the centers occurred in local suboptimal solutions, in order to reduce the size of the new search space, we therefore select a centre into the new search space, only if it is a frequent centre. DEFINITION 5 (Frequent Centre). Given an occurrence threshold β, the local suboptimal clustering result $Z = V^1, ..., V^M$ and the *AB* $a_b one(Z)$, for each centre $v_i \in UZ\ a_b one(Z)$, if the occurrence frequency of v_i exceeds β, then v_i is a frequent centre, where is a set difference operator and $UZ = U_m = 1^M V^m$.

We use the example shown in Table 1 to explain the new search space refinement. From table 1, we find that there are 14 objects be selected as the centre of clusters in V^1, V^2, V^3. The AB of this example is $a_b one(V^1, V^2, V^3) =$ 109, 128, 130, 262, 252, 412, 432, if the occurrence frequency of v_i exceeds β, then it is a frequent centre and will be included in *S*. According to the big valley phenomenon, here we set the parameter $\beta = M * 80\% = 2.4$ (for this example *M*=3), and we obtain the three additional centers 128, 262, 412.

7.3.2.2 Framework of HC_AB

Algorithm 7.5 shows the framework of HC_AB. It works as follows. A K-centre clustering algorithm is run on D with different initializations to generate clustering results Z. The AB a_bone(Z) is generated based on the Definition 2, as shown in step 2. The aim of step 3 is to refine the new search space S according to the description in afore section. The best_V is produced by re-running the K-centre clustering algorithm on S with AB as a fraction of initializations along with the complementary frequent centres.

Algorithm 7.5: HC_AB

Input: D, K, M

Output: *best _V*
  (1) Generate M clustering result, $Z = \{V^1, V^2,..., V^M\}$, by running K-centre clustering algorithm;
  (2) Find the AB $a\_bone(V^1, V^2,..., V^M)$;
  (3) Reconstruct a new search space S;
  (4) Rerun K-centre clustering algorithm with the $a\_bone(V^1, V^2,..., V^M)$ as a fraction initialization centres on S ;
  (5) Return *best _V*.

According to Definition 2, each centre of Z is uniquely represented by a data object ID number in D. For some specific kinds of K-centre clustering algorithms, such as K-means where the centres of clusters algorithms are determined by the mean of all data objects in the cluster, we amend the intersection operator of finding AB described above—the AB is defined as the co-occurrence data objects in the same cluster, and then use their means as the fraction of initialization of K-means clustering algorithm. Heuristic clustering algorithms are sensitive to the initialization problem and are prone to reach the local suboptimal solutions. Due to the strength of AB on improving the performance of heuristic algorithms, many research efforts have introduced it in heuristic algorithm design. In this chapter, we have proposed a novel solution to this by devising an approximate backbone based K-centre clustering approach. The main strength of the proposed method is the capability of restricting the initial search space around the global optimal results by using the approximate backbone, and in turn, reducing the impact of initialization on clustering and improving the efficiency of heuristic clustering. Experiments on several synthetic and real world data sets have shown that the approximate backbone has significant effects on improving the quality of clustering and reducing the initialization impact.

7.4 Improving Clustering Quality in High Dimensional Space

7.4.1 Overview of High Dimensional Clustering

Clustering is one of the frequently used tools in data mining. In many applications, data objects to be clustered are described by points in a high dimensional space, where each dimension corresponds to an attribute/ feature. A distance measurement between any two points is used to measure their similarity. The research in [25] has shown that the increasing dimensionality results in the loss of contrast in distances between data objects. Thus, clustering algorithms that measure the similarity between data objects based on all attributes/features of the data tend to degrade in high dimensional data spaces. In addition, the widely used distance measurement usually perform effectively only on some particular subsets of attributes, where the data objects are distributed densely [20]. In other words, it is more likely to form dense and reasonable clusters of data objects in a lower dimensional subspace [1]. Recently, several algorithms for discovering clusters of data objects in subsets of attributes have been proposed, and they can be classified into two categories: subspace clustering and projective clustering [32]. Subspace clustering was first proposed by Agrawal in [1]. The main task of subspace clustering is to search clusters in 2d subspaces of a data set according to their individual cluster definition. A large number of overlapping clusters are typically reported. Most of the cluster definitions of subspace clustering are based on a global density threshold that ensures anti-monotonic properties necessary for an Apriori style search. The setting of global density threshold heavily relies on the domain knowledge and has a significant impact on clustering results. Large values of the global density threshold will result in only low dimensional clusters, whereas small values will lead to not only higher dimensional clusters but also a large number of low dimensional clusters (many of which are too trivial to be kept) [32]. CLIQUE [1], ENCLUS [11], SSC [40] and SCUD [12] are the typical subspace clustering algorithms in the lecture. In general, subspace clustering aims to find out overlapped clustering results in a bottom-up way, while, projective clustering seeks to assign each point to a unique cluster (clusters embedded in different subspaces) in a top-down way. PROCLUS [9] is one of the classical projective clustering algorithms. It discovers groups of data objects located closely in each of the related dimensions in its associated subspace. In such case, the data objects would spread along certain directions which are parallel to the original data axes. ORCLUS [10] aims to detect arbitrarily oriented subspaces formed by any set of orthogonal vectors. EPCH [34] is focused on uncovering projective clusters with varying dimensionality, without requiring users to input the expected average dimensionality l of the associated subspace and the number of clusters K that inherently exists in the data set. The d-dimensional histogram created with equal width, is used to capture the dense units and their locations in the d-dimensional space. A compression structure is used to store these dense units and their locations. At last, a search method is used to merge similar and adjacent dense units and form subspace clusters. P3C [31] can effectively discover projective clusters in the data while minimizing the number of required parameters. P3C also does not need the number of projective clusters as input and can discover the true number of clusters. There are three steps consisted in P3C. Firstly, regions corresponding to the clusters on each attribute are discovered. Secondly, a cluster core structure described by a combination of the detected regions is designed to capture the dense areas in a high dimensional space. Thirdly, cluster cores are refined into projective clusters, outliers are identified, and the relevant attributes for each cluster are determined. STATPC [32] uses a varying width hyper-rectangle structure to find out the dense areas embedded in the high dimensional space. By using a spatial statistical method, all dense hyper-rectangles are found. A heuristic search process is run to merge these dense hyper-rectangles and clustering results are generated. The clusters of projective clustering are defined as the dense areas in corresponding subsets of attributes. In projective clustering, it is a common way that a hyper-rectangle structure is used to find out the dense areas in the d-dimensional space at first; and then, a search method is run to merge these hyper-rectangles for generating clusters. Because the dense area is captured by the hyper-rectangle structure, it is important to define the structure before clustering. There are two kinds of hyper-rectangle structures used in projective clustering—the equal width hyper-rectangle and the varying width hyper-rectangle. For the equal width hyper-rectangle structure, each dimension is divided into equal width intervals, and the hyper-rectangles are constructed by these intervals, for instance, the d-dimensional histogram is used as the first step of the construction of hyper-rectangle structure in EPCH. As for the varying hyper-rectangle structure, it (1) randomly selects a data object from a data set D; (2) constructs a hyper-rectangle structure around the data object with randomly selected widths; (3) runs a statistical test on the hyper-rectangle to decide whether it is a dense hyper-rectangle. In real applications, it is a difficult task to set reasonable widths for these hyper-rectangles.

7.4.2 Motivation of our Method

Furthermore, data objects may belong to various clusters in different subspaces. Projective clustering is an efficient way to deal with high dimensional clustering problems. Explicitly or implicitly, projective clustering algorithms assume the following definition: Give a data set D of d-dimensional data objects, a projected cluster is defined as a pair (C_k, S_k), where C_k is a subset of data objects and S_k is a subset of attributes such that the data objects in C_k are projected along each attribute in S_k onto a small range of values, compared to the range of values of the whole data set in S_k, and the data objects in C_k are uniformly distributed along every other attributes not in S_k. The task of projective clustering is to search and report all projected clusters in the search space. Generally, researchers often define the equal or varying width hyper-rectangle structure to capture the dense area at first, and then merge these dense areas to generate the projected clusters. In real applications, however, it is hard to decide the widths of these hyper-structures directly. The use of a histogram is a common and easy way to define the width of hyper-rectangles. The width defined by this method has a strict bond of distribution which heavily affects the quality of the clustering result. On the other hand, it is intuitive to decide the width directly from the density distribution estimated from the real data itself. Kernel width of a data object derived by using kernel estimator has the ability of capturing the dense distribution around it [43]. So it is a wise way that use kernel estimator to decide the width of hyper-rectangle structure. Conversional kernel estimators, however, can not deal with high dimensional data. Rodeo [27]. An efficient local kernel estimator, has the ability of estimating the kernel width around a data object in the high dimensional space. By making use of Rodeo, we propose an innovative projective clustering in this paper. Particularly, in this paper, we define a new structure named Significant Local Dense Area (SLDA) to capture the local dense area around the data object based on Rodeo and spatial statistical theory; and then propose a greedy search algorithm to generate whole SLDAs which could cover the data distribution in the d-dimensional space; eventually, we merge the SLDAs to construct the projected clusters and filter out outliers.

7.4.3 Significant Local Dense Area

7.4.3.1 Rodeo Algorithm

Let $x_1, x_2, ..., x_N, x_i \in R^d (i = 1, ..., N)$ be a sample set from a distribution F with density function f. Non-parametric density estimation methods are often used to estimate the f. Rodeo is an effective kernel density estimator for sparse and high dimensional data and it has the advantages of performing the selection of subset of attributes and the determination of kernel width simultaneously [43].

The kernel density estimator is defined as

$$\widehat{f}w(x) = \frac{1}{N \det(W)} \sum_{i=1}^{N} K(W^{-1}(x - X_i))$$

where $K(.)$ is a symmetric kernel with $\int K(u)du = 1$ and $\int K(u)du = 0$, and $W = diag(width_1, \ldots, width_d)$ is a diagonal matrix. Rodeo algorithm uses an iterative learning step to calculate the kernel value and estimate the kernel width W of data object x. In order to reduce the time consuming of Rodeo, the authors propose a greedy Rodeo method by embedding the Rodeo within LARS (Least Angle Regression) [27]. In this chapter, we prefer to select this modified Rodeo as our subroutine in our experiments. Here we assume that the attributes which have the significant contribution on estimating the kernel value of x are listed in the first r columns in W and the rest attributes are irrelative with the kernel estimation. So the kernel width W returned by Rodeo satisfies the following Theorem [27].

Theorem 7.1. Given the kernel width $W = diag(width_1, \ldots, width_d)$, it satisfies:

$P(w_j = w_j(0)) \to 1$ for all $j > p$ and $P(w^0_j (Nb_N)^{\frac{-1}{4+r}}) \le w_j \le w^0_j (Na_N)^{\frac{-1}{4+r}}) \to 1$, where w^0_j is the original large kernel width on attribute j, a_N and b_N are the constant.

From Theorem 7.1, we can find that the kernel widths on relative attributes which have the significant contribution to the kernel estimation are smaller than those original kernel widths. In this paper, we therefore select these r attributes to construct the relative subspace S and its corresponding kernel width set W^* to form the hyper-rectangle.

In order to show the ability of Rodeo on finding out the subset of attributes that has significant efforts on the kernel estimation and the corresponding kernel width determination, we give an example. Randomly generate a data set D containing 50 data objects, each object described by 4 attributes, the range of each attribute is in [0,1]. The data of the first two attributes is subject to a normal distribution and the data of the last two attributes is subject to a uniform distribution, as shown in Fig. 7.4.1(a) and (b). Randomly select a data object x from D, and use the Rodeo method to estimate the kernel density of it. The subset of attributes which has significant contributions to the kernel density estimation is on dimension 1 and 2. The kernel widths of x are shown in Fig. 7.4.1(c). From Fig. 7.4.1(c), we can observe that the kernel width on dimension 1, 2 is smaller than the widths on dimension 3, 4, which indicates the fact that the contribution of dimension 1, 2 is higher than that of dimension 3, 4 to the kernel density estimation.

7.4.3.2 Definition of SLDA

Let $D = x_1, x_2, ..., x_N$ be a data set of N d-dimensional data objects. Let $A = attr_1, attr_2, ..., attr_d$ be the set of attributes of the data objects so that $xj \in dom(attr_j)$, where $dom(attr_j)$ denotes the domain of the attribute $attr_j, j = 1, ..., d$. Without losing the generality, we assume that all the attributes have been normalized, i.e., $dom(attr_j) \in [0, 1]$. In this section, we first construct a hyper-rectangle structure of data object x based on the kernel width W^* and determine the corresponding subset of attributes S by running Rodeo; and then, conduct a spatial statistical test on the hyper-rectangle to decide whether it is a SLDA around the data object x. To better describe the process of determining the SLDA, we introduce the following definitions.

Definition 7.6 (Hyper-Rectangle). Given a subset of attributes S and a kernel width $W^* = width_1, ...,width_p$ of data object x, the hyper-rectangle structure H around x can be constructed as : $H = I_1???I_p$, where $I_j = [x_j?width_j/2, x_j +width_j/2], j = 1, p, p = dim(S)$.

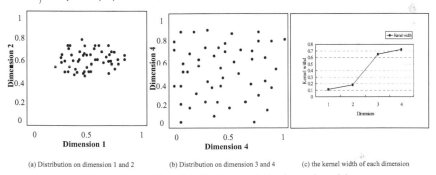

(a) Distribution on dimension 1 and 2   (b) Distribution on dimension 3 and 4   (c) the kernel width of each dimension

Figure 7.4.1: The distribution and the kernel width

The volume of hyper-rectangle around the data object x is $vol(H) = \Pi_{j=1}^p width_j$. The number of data objects within the hyper-rectangle H can be used to indicate the local density around x.

Definition 7.6 (Data Objects in Hyper-Rectangle). Given a data set D and a hyper-rectangle H around a data object x in the subset of attributes S, the data objects located in H are defined as: $remm(H) = x_i \in D \mid (x_j - width_j/2) \leq xj \leq (x_j - width_j/2), i = 1, ...,N; j = 1, ..., p$.

Definition 7.7 (Local Density of Hyper-Rectangle). Given the identified data object set remm(H) of a data object x , the local density around x in the subset of attributes S is defined as: $LS(H) = \mid remm(H) \mid$, where $\mid \cdot \mid$ means the number of objects within the hyper-rectangle.

Known from the spatial statistical theory of assigning N data points in a space, the number of data points which are assigned in a bounded area is subject to the Binomial distribution with parameters of N and the volume of the bound area [16]. Here H is the hyper-rectangle in the subset of attributes, which is a bounded area, and the local density LS(H) is the number of data objects assigned in H. So $LS(H)$ is subject to the Binomial distribution with parameter N and vol(H), i.e.,

   $LS(H) \lozenge Binomial(N, vol(H))$.

To decide whether H is a dense hyper-rectangle, we run a null hypothesis statistical test on H.

   h_0: Hyper-rectangle H in S contains $LS(H)$ data objects.

The significant level α of the statistical hypothesis is a fixed probability of wrongly rejecting the null hypothesis, when in fact it is true. α is also called the rate of false positives or the probability of type I error. The critical value of the statistical hypothesis test is a threshold which the value of the test statistic is compared to determine whether the null hypothesis is rejected. There are two test methods for hypothesis test: one-side test and two-side test. For the one-side test, the critical value θ is computed based on $\alpha = p(LS(H) > \theta)$. For the two-side test, the computation of left critical value θ^L is the same as the one-side test, but the right critical value θ^R is computed based on $\alpha = p(LS(H) > \theta^R)$ Where is a probability function.

Definition 7.8 (Significant Local Dense Area). Let H be a hyper-rectangle in the subset of attributes S. Let α be a significant level and θ be the critical value computed at the significant level α based on the one-side test, where the probability is computed using $Binomial(N, vol(H))$. If $LS(H) > \theta$, (H, S) is deemed as a Significant Local Density Area (SLDA) around the data object x.

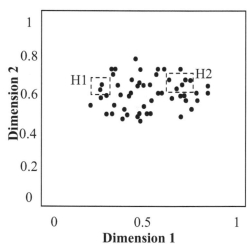

Figure 7.4.2: Hyper-rectangle of dataset D

7.4.4 Projective Clustering based on SLDAs

7.4.4.1 Finding SLDAs

After using the Rodeo method, the subset of attributes S and the kernel width w^* of a data object x are generated. Based on S and w^*, a hyper-rectangle structure H is constructed around x. We also use the data set described in Section 2.2 as an example to illustrate the constructed hyper-rectangle. The subset of attributes having important supports on kernel estimation is on dimensions 1 and 2.

The example of H is shown in Fig. 7.4.2. In this figure, there are two hyper-rectangles H_1, H_2 around two randomly selected data objects. But the local density values of each hyper-rectangle are different. Based on the Definitions 7.6 and 7.7, the local density value of the dashed frame (H_1) is 2, whereas the local density value of the real line frame (H_2) is 5. According to the spatial statistical hypothesis described in Section 2.3, we set the significant level $\alpha = 0.001$ and the value of θ calculated by the one-side statistical test is 3.21. So the dashed frame is not a significant local density area as $LS(H_1) < \theta$, and it will be deleted.

The greedy search method, G_SLDA as shown in Algorithm 7.6, is utilized to find the whole significant local dense areas which can cover the data distribution of D in the d-dimensional space. The main steps of G_SLDA are to:

(1) randomly select a data object x from D and calculate the subset of attributes S and the kernel width W^* using Rodeo (steps 3–4);
(2) create a hyper-rectangle structure H around x based on Definition 7.5, and obtain the local dense value of H based on Definition 7.6 and 7.7 (steps 5–6);

Algorithm 7.6: G_SLAD

Input: Data set D

Output: *SLDA* set *SLD*

1. *SLD* $\varnothing$;
2. Loop
3.   Randomly select a data object x from D;
4. $[S,W^*]$ =Rodeo(D, x);
5. H = CreateH(x, S, W^*);
6. $remm(H) = $ G_remm(D, H);
7. Run a statistical test of $LS(H) = | remm(H) |$ on H according to definition 4;
8.   If $LS(H) \geq \theta$
9. $SLD = SLD \cup (H, S)$;
10. x and the data object in $remm(H)$ are signed as visited;
11.   end
12. Until all data objects in D are visited
13. return *SLD*

---

(3)   decide whether *H* is a Significant Local Dense Area based on Definition 7.8. If *H* is a significant local dense area, store the pair (*H, S*) into *SLD* set, and label *x* and the data objects in *H* as visited, otherwise, discard the obtained hyper-rectangle *H* (steps 7–11). These steps iteratively run until all data objects have been processed.

7.4.4.2 Generating Clusters by Merging SLDAs

G_SLDA find out all the significant local density areas of data set *D*. Each SLDA, (*H, S*), in *SLD* contain a density hyper-rectangle satisfying $LS(H) \geq \theta$ and its relevant subset of attribute *S*. The main structure of the density area in a subset of attributes can be captured by all the density hyper-rectangles embedded in it. To further understand the relationship between the density area and significant local density area, we give an example to show it. Given a data set *D* used in Fig. 7.4.1, each object is described by four attributes, the range of each attribute is in [0,1]. The data of the first two attributes is subject to a normal distribution and the data of the last two attributes is subject to a uniform distribution.

From the above discussion, we know that the identified subset of attributes is on *S* = 1, 2 denoted by nodes in Fig. 7.4.2, and the SLDAs generated by G_SLDA on *S* = 1, 2 are represented by six solid rectangles (i.e., H_1, H_6). From Fig. 7.4.3, we therefore can conclude that the main structure of data distribution is characterized by six dense hyper-rectangles.

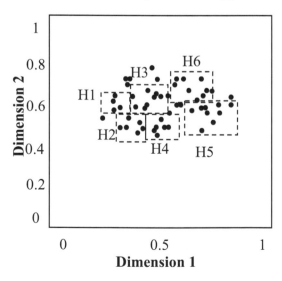

Figure 7.4.3: Example of the relationship between SLDAs and the density area

The clustering result of projective clustering is represented by the data objects which are located in small ranges in its specific subspace. As for SLDA, the various *Hs* do indicate these small ranges in *S*. The data objects that are scattered in deferent hyper-rectangle *Hs* but within the same subspace *S* constitute the different parts of a dense area in *S*. To obtain the fully projected clusters, we need to merge these hyper-rectangles within the same subspace *S* to generate its corresponding cluster. More concretely, the clusters derived by a projective clustering algorithm are represented by all the dense areas and their related subsets of attributes.

Algorithm 7.7: *MC_SLDA*

Input. *SLD*

Output. Clustering results and outliers
(1) Divide *SLD* into several subsets;
(2) For each subset of *SLD*, a single-linkage merger algorithm is run to find out the clustering results;
(3) Refine the clustering results

Hence we merge all the dense hyper-rectangles in the same subset of attributes to generate the clusters. Further, for the example shown in Fig. 7.4.3, the cluster in the subset of attributes $S = 1, 2$ is $(\cup_h = 1^6, (1, 2))$, where $\cup_h = 1^6$ denotes the merger processing of density hyper-rectangles. The three major steps of the merger clustering algorithm on SLDAs, named MC_SLDA, are to (1) divide SLDAs into several subsets so that the hyper-rectangles within one attribute subset have the same subspace S; (2) run a single-linkage merger algorithm on these subsets to find the fully projected significant local dense area; (3) refine the clustering results and detecting the outliers. The pseudo codes of *MC_SLDA* are detailed in Algorithm 2. The data objects which are not included in any clusters are denoted as *Rest* $= D \setminus (\cup_{k=1}^{K} C_k)$,
where $\setminus$
is the set different operator. In the clustering result refinement, we use the reassign method proposed in [39] to assign data objects in Rest to the corresponding clusters. After the refinement, the data objects which do not belong to any clusters can be regarded as outliers, and an outlier collection is generated.

Hyper-rectangle structure is often used in finding the density area in high dimensional data sets. The determination of the width of hyper-rectangle structure is a crucial task in high dimension clustering applications. The majority of projective clustering algorithms use the restrictive model to determine the width of hyper-rectangle, which has significant efforts on discovering real clustering results. Inspired by the kernel density method, we present a new way to design the hyper-rectangle structure, whose width is determined by the true data distribution. In order to examine whether a hyper-rectangle structure is a Significant Local Density Area, we run a spatial statistical test on it. A greedy algorithm is proposed to find out all the SLDAs in the data set. At last, a merger algorithm is applied on SLDAs to generate the clustering results and identify the outliers. Experiment results on synthetic and real data sets have shown that our method outweighs the traditional projective clustering algorithms in discovering the high quality clustering results with varying density widths.

7.5 Chapter Summary

In this chapter, we focused on the modification and improvement of heuristic clustering algorithms at first. Two different methods based on smoothing space and approximate backbone respectively, were proposed to deal with the drawbacks of heuristic clustering algorithm. For the method based on smoothing space, different smoothing operator to reconstruct the search space of clustering is used. The search space construction step needs more executing time. For the method based on approximate, different clustering results which are derived by other heuristic clustering algorithm, such as k-means, VSH, etc., to capture the common part of the dataset are used. It also needs more executing time for generating the clustering results. And then, we discussed the method to improve the quality of projective clustering algorithm in high dimensional space. A method based on Rodeo and statistical theory was been proposed in this chapter. This method has the ability to capture the real distribution of data objects in high dimensional space. However, the Rodeo method needs more running time. It is an open question that how to reduce the time cost of these three methods.

References

[1] A. Hinneburg, C. C. Aggarwal and D. A. Keim. Automatic subspace clustering of high dimensional data for data mining applications. *Sigmod Record*, 27(2): 94–105.
[2] D. Arthur and S. Vassilvitskii. k-means++: the advantages of careful seeding. *In: Proceedings of the eighteenth annual ACM-SIAM symposium on Discrete algorithms Algorithms*, pp.: 1027C1035, 2007.
[3] B. Addis, M. Locatelli and F. Schoen. Local optima smoothing for global optimization. *Optimization Methods and Software*, 20(4-5): 417–437, 2005.
[4] M. A. Belal. A New Algorithm for Cluster Initialization. *Enformatika Conferences*, pp. 74–76, 2005.
[5] J. F. Brendan and D. Dueck. Clustering by Passing Messages Between Data Points. *Science.*, 5814(315): 972–976, 2007.
[6] P. S. Bradley and U. M. Fayyad. Refining Initial Points for K-Means Clustering. *In: Proceediing of International Conference on Machine Learning*, pp. 91–99, 1998.
[7] Brendan J. Frey and Delbert Dueck. Response to Comment on "Clustering by Passing Messages Between Data Points". *Science*, 319(5864): 726, 2008.

[8]   A. Caprara and G. Lancia. Optimal and Near—Optimal Solutions for 3D Structure Comparisons. *In: Proceeding of the First International Symposium on 3D Data Processing Visualization and Transmission*, pp. 734–744, 2002.

[9]   C. C. Aggarwal, C. M. Procopiuc, J. L. Wolf and P. S. Yu. Fast Algorithms for Projected Clustering. *Sigmod Record*, 28(2): 61–72, 1999.

[10] C. C. Aggarwal and P. S. Yu. Finding generalized projected clusters in high dimensional spaces. *Sigmod Record*, 70–81, 2000.

[11] Chun-Hung Cheng, Ada Wai-Chee Fu and Yi Zhang. Entropy-based subspace clustering for mining numerical data. *In: Proceedings of the International Conference on Knowledge Discovery and Data Mining*, pp. 84–93, 1999.

[12] E. Ehsan and R. Vidal. Sparse subspace clustering. *Computer Vision and Pattern Recognition*, 2790–2797, 2009.

[13] P. Drineas, A. M. Frieze, R. K., S.Vempala and V. Vinay. Clustering Large Graphs via the Singular Value Decomposition. *Mach. Learn*, 56(1-3): 9–33, 2004.

[14] A. Forgy, N. Laird and D. Rubin Maximum likelihood from incomplete data via codes for analog input patterns. *Applied Optics*, (26): 4919–4930, 2004.

[15] M. R. Garey, D. S. Johnson and H. S. Witsenhausen. The complexity of the generalized Lloyd—Max problem. *IEEE Transactions on Information Theory*, 28(2): 255–256, 1982.

[16] J. Gu and X. F. Huang. Efficient local search with search space smoothing: a case study of the traveling salesman problem (TSP). *IEEE Transactions on Systems, Man, and Cybernetics*, 24(5): 728–735, 1994.

[17] J. He, M.Lan, C. L. Tan, S. Y. Sung and H. B. Low. Initialization of cluster refinement algorithms: a review and comparative study. *In: Proceeding of International Symposium on Neural Networks*, pp. 297–302, 1998.

[18] A. K. Jain and R. C. Dubes. Algorithms for Clustering Data, Prentice Hall College Div, 1988.

[19] H. Jiang, X. C. Zhang, G. L. Chen and M. C. Li. Backbone analysis and algorithm design for the quadratic assignment problem. *Science in China Series F: Information Sciences*, 51(5): 476–488, 2008.

[20] S. Jeffrey, Beis and David G. Lowe. What is the Nearest Neighbour in high dimensional spaces. *In: Proceeding of the 26th International Conference on Very Large Databases*, pp. 506–515, 2000.

[21] T. J. Tou and R. C. Gonzalez. Pattern recognition principles. *Addison-Wesley Pub. Co.*, 1974.

[22] I. Katsavounidis, C. C. Kuo and Z. Zhang. A new initialization technique for generalized Lloyd iteration *IEEE Signal Processing Letters*, 1(10): 144–146, 1994.

[23] L. Kaufman and P. J. Rousseeuw. Finding Groups in Data: An Introduction to Cluster Analysis *Wiley-Interscience*, 2005.

[24] K. D. Boese Cost Versus Distance In the Traveling Salesman Problem. *Wiley- Interscience*, 1995.

[25] S. B. Kevin, J. Goldstein, R. Ramakrishnan and U. Shaft. When Is "Nearest Neighbor" Meaningful? *In: Proceeding of International Conference on Database Theory*, pp. 217–235, 1999.

[26] S. Kirkpatrick and G. Toulouse. Configuration space analysis of travelling salesman problems. *Journal De Physique*, 46(8): 1277–1292, 1985.

[27] J. D. Lafferty and L. A. Wasserman. Rodeo: Sparse Nonparametric Regression in High Dimensions. *In: Proceeding of the International Conference on Neural Information Processing Systems*, pp. 1–32, 2007.

[28] J. F. Lu, J. B. Tang, Z. M. Tang and J. Y. Yang. Hierarchical initialization approach for K-Means clustering *Pattern Recognition Letters*, 29(6): 787–795, 2007.

[29] J. MacQueen. Some methods for classification and analysis of multivariate observations *Berkeley Symposium on Mathematical Statistics and Probability*, (1): 281–297, 1967.

[30] P. Merz and B. Freisleben. Fitness Landscapes, Memetic Algorithms and Greedy Operators for Graph Bi-Partitioning. *Evolutionary Computation*, 8(1): 61–91, 2000.

[31] G. Moise, J. Sander and M. Ester. P3C: A Robust Projected Clustering Algorithm. *In: Proceeding of IEEE International Conference on Data Mining*, pp. 414–425, 2006.

[32] G. Moise and J. Sander. Finding non-redundant, statistically significant regions in high dimensional data: a novel approach to projected and subspace clustering. *Knowledge Discovery and Data Mining*, 533–541, 2008.

[33] K. Niu, S. B. Zhang and J. L. Chen. An Initializing Cluster Centers Algorithm Based on Pointer Ring. *In: Proceeding of the Sixth International Conference on Intelligent Systems Design and Applications*, pp. 655–660, 2006.

[34] Ka Ka Ng Eric, Ada Wai-chee Fu and Raymond Chi-wing Wong. Projective Clustering by Histograms. *IEEE Transactions on Knowledge and Data Engineering*, 17(3): 369–383, 2005.

[35] C. R. Reeves. Landscapes, Operators and Heuristic Search. *Annals of Operations Research.*, 86(0): 473–490, 2000.

[36] S. Sahni and T. F. Gonzalez. P-Complete Approximation Problems. *Journal of the ACM*, 23(3): 555–565, 1976.

[37] SAS Institute SAS User's Guide Statistics Version 5. *Sas Inst*, 1985.

[38] C. Sharlee and W. X. Zhang. Searching for backbones and fat: A limit-crossing approach with applications. *AAAI/IAAI'02*, pp. 707–712, 2002.

[39] Stephen J. Redmond and Conor Heneghan. A method for initialising the K-means clustering algorithm using kd-trees. *Pattern Recognition Letters*, 28(8): 965–73, 2007.

[40] G. Stephan, H. Kremer and T. Seidl. Subspace Clustering for Uncertain Data. *SIAM Inter- national Conference on Data Mining*, pp. 385–396, 2010.

[41] Kuo-lung Wu and Miin-shen Yang. Mean shift-based clustering. *Pattern Recognition*, (40): 3035–3052, 2007.

[42] W. X. Zhang. Phase Transitions and Backbones of 3SAT and Maximum 3SAT. *Principles and Practice of Constraint Programming*, 153–167, 2001.

[43] Z. H. Zhou and W. Tang. Clusterer ensemble. *Knowledge Based Systems*, 19(1): 77–83, 2006.

[44] P. Zou, Z. Zhou, G. L. Chen, H. Jiang and J. Gu. Approximate-Backbone Guided Fast Ant Algorithms to QAP. *Journal of Software*, 16(10): 1691–1698, 2005.

CHAPTER 8

Multi-Label Classification

8.1 Introduction

With the advent of affordable digital facilities and the rapid development of internet, a huge amount of data has been generated, and the volume is still increasing explosively. In most cases, such data needs to be analysed and thus potential knowledge hidden behind them could be extracted and summarized. Classification, also known as supervised learning, is a fundamental technique for data analysis. In the framework of classification, each object is described as an instance, which is usually a feature vector that characterizes the object from different aspects. Moreover, each instance is associated with one or more labels indicating its categories. Generally speaking, the process of classification consists of two main steps: the first is training a classifier based on a set of labelled instances, the second is using the classifier to predict the label of unseen instance.

It is usually assumed that each instance has only one label. Let X denote the instance space and L be a set of labels, a single-label training set could then be denoted as $D = \{(x_1, l_1), (x_2, l_2), \ldots, (x_n, l_n)\}$, where x_i is the ith instance and l_i is its relevant label taken from L. The objective of classification could be viewed as learning a mapping from the instance space to the label space: $f : X \rightarrow Y$, based on a training dataset D. Generally, this kind of learning is also called single-label classification [1].

However, the instances might be assigned with multiple labels simultaneously, and problems of this type are ubiquitous in many modern applications. For instance, Fig. 8.1.1 gives us two examples of multi-label objects. As it has shown, a film could be tagged as *action*, and *adventure* simultaneously, and an outdoor scene can also be viewed as a *sea* scene, etc. Learning for this kind of instances is called multi-label classification accordingly.

Recently, there have been a considerable amount of research concerned with dealing with multi-label problems and many state-of-the-art methods has already been proposed. It has also been applied to lots of practical applications, including text classification [2, 3], gene function prediction [4], music emotion analysis [5], semantic annotation of video [6], tag recommendation [7, 8] and so forth. Motivated by the increasing applications, multi-label classification is becoming a hotspot and attracting more and more academic researchers and industrial practitioners.

In this chapter, a comprehensive and systematic study of multi-label classification is carried out, in order to give a clear description of what is the multi-label classification and highlight the basic and representative methods. The chapter is organized as follows. First, we introduce the formal concept of multi-label classification and several definitions from different perspectives. Second, we divide the multi-label methods into several types, and provide a thorough description of the state-of-the-art methods according to different types. Subsequently, we present a number of frequently-used benchmark datasets and evaluation metrics, which are used to examine and compare the performance of different methods. Finally, the conclusions and the further challenges are given in the last section.

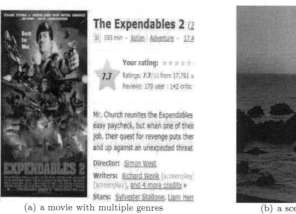

(a) a movie with multiple genres   (b) a scene have several meanings

Figure 8.1.1: Two examples of multi-label instances

8.2 What is Multi-label Classification

To begin with, let us give the formal concept of multi-label classification firstly, in order to gain a better comprehension of it, and make the analysis and comparison of the following algorithms more easy.

Let X denote the instance space and $L = \{l_1, l_2, ..., l_m\}$ be a set of labels, so $D = \{(x_1, C_1), (x_2, C_2), \ldots, (x_n, C_n)\}$ can be used to denote a set of multi-label instances, where $xk \in X$ is an instance, and $C_k \subset L$ is a subset of L that denote $x_{k's}$ true labels. The target of multi-label classification is thus to learn a classifier: $f : X \to 2^L$, which is a mapping from the instance space to the space consists of all the possible subsets of L, where 2^L is the power set of L. Ck can also be represented as a Boolean vector $y_k = (b_{k1}, b_{k2}, \ldots, b_{km})$, where $b_{kj} = 1$ indicates label l_j is $x_{k's}$ true label ($l_j \in C_k$), whereas $b_{kj} = 0$ indicates the opposite.

In the view of probability, the process of multi-label classification can also be viewed as to calculate the posteriori joint probability of each possible label vector. If x is an unlabelled instance, and Y is a set of possible label vector, then the reasonable label vector, for instance, x should be the one that gets the greatest posteriori probability, as illustrated by Formula 8.2.1.

$$y_x = \arg\max_{Y=y} P(Y \mid x) \qquad (8.2.1)$$

However, it's a NP-hard problem to calculate the posteriori probability by using Formula 8.2.1 directly, so it's very difficult and time-consuming. One practicable alternative is to learn the relationship between labels and characterize the joint probability as a Bayesian network. For each label y_k, let $Parent(l_k)$ denote the set of labels that label l_k is dependent on, so $P(y \mid x)$ can be transformed as Formula 8.2.2

$$P(y \mid x) = \prod_{k=1}^{m} P(y_k \mid parent(y_k), x) \qquad (8.2.2)$$

where y_k denotes the kth label. Hence we can get the label vector's posterior probability by computing each label's posterior probability respectively and then multiplying them. It's clear that how to find the dependent labels for each label is emerging as a critical issue now. Actually, how to learn the appropriate dependencies among labels in order to improve the learning performance is getting more and more important. We can also see various methods in the following sections that try to incorporate label dependencies into the learning process, and most of them indeed bring benefit.

The predictions can also be a number of values besides Boolean label vector in many cases, and each value indicates a label's probability or confidence being the instance's true label. In these cases, the task of multi-label classification could be transformed to learning a function $f : X \times L \rightarrow R$, and thus $f(x, l)$ outputs a real value that indicates label l's confidence to be instance x's true label. Actually, many multi-label algorithms learn such a function for each label, in other words, $f = \{f_1, f_2, \ldots, f_m\}$. Thus the predictions for instance x takes the form as shown in Formula 8.2.3.

$$f(x) = \{f_1(x), f_2(x), \ldots, f_m(x)\} \qquad (8.2.3)$$

The multi-label classification problem thus can be tackled through solving a label ranking (LR) problem, where the labels are sorted according to their predicted values in descending order, then a threshold t, is learned to decide the relevant labels and irrelevant labels. For label l_i, it is predicted as a relevant label if $f_i(x) \geq t$, otherwise it is predicted as an irrelevant label. LR is a very important issue in multi-label learning, not only it is an effective means of solving multi-label problems, but also it's more suitable for some real applications. For example, users of tag recommendation system might prefer seeing the most interesting tags top the tag list, instead of just a set of tags. So it poses a very interesting and useful way to solving multi-label classification. Usually, LR models can also be learned for single-label instances to solve the multi-class problems.

Nowadays, researchers have proposed lots of effective and efficient methods for multi-label classification, and they mainly fall into two main categories: (1) algorithm adaptation (2) problem transformation [9]. Algorithm adaptation methods extend traditional single-label algorithms, such as kNN, decision tree, Naive Bayes etc., in order to enable them to handle multi-label data directly [4, 10, 11]. By contrast, problem transformation methods decompose the multi-label instances into one or several single-label instances, thus existing methods could be used without modification [12, 13, 14]. In other words, the algorithm adaptation strategy is to fit the algorithms to data, whereas the problem transformation strategy is to fit data to the algorithms. The primary difference between these two strategies is that the algorithm adaptation is algorithm-specific, therefore the strategy used in one method can not be applied to another one usually. Nevertheless the problem transformation is algorithm-independent, so it is more flexible and can be used with any existing models. The following sections will elaborate on them by different categories.

8.3 Problem Transformation

As mentioned above, problem transformation is a fundamental strategy for tackling multi-label problems. It enables most of the existing methods to work easily, while making few modifications to them. Hence it gets much popularity among researchers, and various methods based on it have been proposed [12, 13, 14, 15, 16]. Several simple methods of this kind are All Label *Assignment* (ALA), No Label Assignment (NLA), Largest Label Assignment (LLA) and Smallest Label Assignment (SLA), as summarized by Chen et al. and used for multi-label document transformation [17]. Let's explain these methods through a multi-label dataset shown in Table 8.1.

Table 8.1: A exmaple of multi-label dataset

instances	labels
x_1	l_1, l_3
x_2	l_1, l_2, l_4
x_3	l_2
x_4	l_2, l_4

Suppose that we have a set of instances $\{x_1, x_2, x_3, x_4\}$, as shown in Table 8.1. For each instance, ALA approach will make a copy of it and assign the copy to all its true labels respectively. Namely, ALA will replace an instance $\{x_i, C_i\}$ with $|C_i|$ instances, each of which is associated with one label in C_i. Thus instance x_i will appear $|C_i|$ times in the dataset. The NLA approach only keeps instances, with a single label and discards others with more than one label. Different from the previous two approaches that either keep or delete all the labels of a multi-label instance, LLA and SLA make a compromise by keeping only one label as the instance's final label and discarding the remaining labels. LLA selects the label that is most frequent among all instances, whereas SLA selects the label that is least frequent among all instances instead. The results of transformation for the above example dataset using these 4 approaches Chen alsaorepsrhoopwonseidn aFignuorveel8.a3p.1p.roach is called entropy-based weighted label assignment (ELA). This method actually assigns a weight to each instance that is generated by ALA approach [17]. Generally speaking, the aforementioned approaches are straightforward, but they ignore lots of potential information in the discarded instances and labels, which could have been used to enhance the learning performance. Currently, two commonly used strategies for problem transformation are label powerset (LP) method that treats all the true labels of each instance as a new single label, and binary relevance (BR) method that predicts each label respectively [13, 9]. A considerable mount of methods based on these two strategies have been proposed, some of them will be inspected in the following parts respectively.

8.3.1 Binary Relevance and Label Powerset

BR method transforms a multi-label dataset D into $|m|$ single-label datasets $\{D_1, D_2, \ldots, D_m\}$, each one corresponding to a label. Dataset D_i includes the same instances of D, while each instance's label is changed to 1 if l_i is its true labels in D, otherwise it is changed to 0. The generated datasets for dataset given in Table 8.1 through using BR method are shown in Fig. 8.3.2.

insts	label
x_1	l_1
x_1	l_3
x_2	l_1
x_2	l_2
x_2	l_4
x_3	l_2
x_4	l_2
x_4	l_4

(a) ALA's result

insts	label
x_3	l_2

(b) NLA's result

insts	label
x_1	l_1
x_2	l_2
x_3	l_2
x_4	l_2

(c) LLA's result

insts	label
x_1	l_3
x_2	l_1
x_3	l_2
x_4	l_4

(d) SLA's result

Figure 8.3.1: The Transformation results using ALA, NLA, LLA, and SLA

insts	l_1
x_1	1
x_2	1
x_3	0
x_4	0

(a) D_1

insts	l_2
x_1	0
x_2	1
x_3	1
x_4	1

(b) D_2

insts	l_3
x_1	1
x_2	0
x_3	0
x_4	0

(c) D_3

insts	l_4
x_1	0
x_2	1
x_3	0
x_4	1

(d) D_4

Figure 8.3.2: Transformation results through using BR method

We can see that one dataset is generated for each label, thus the original multi-label problem is changed to multiple binary problems. Then a binary classifier f_i can be trained based on dataset D_i, which is responsible for giving a *yes/no* prediction for label l_i. In the end, results from all the binary classifiers will be aggregated into a final prediction. Basic BR approach is very straightforward and efficient, and has become a benchmark algorithm for being compared with other algorithms. However, it assumes that the labels are independent each other and ignores the potential label dependencies during the transformation process.

LP method deals with the multi-label problem from a different perspective. Instead of treating labels respectively, it converts the label set of each instance into a single label. Thus a set of single-label instances is formed, where the label space consists of all the possible subsets of the original label set. After being transformed through using LP method, the result for dataset given in Table 8.1 should be as shown in Table 8.2.

Table 8.2: The transformation results through using basic LP method

instances	labels
x_1	$\{l_1, l_3\}$
x_2	$\{l_1, l_2, l_4\}$
x_3	$\{l_2\}$
x_4	$\{l_2, l_4\}$

We can see that the result of LP method is only one dataset instead of generating one dataset for each label as BR does. Since each instance's new label is a subset of original set of labels, the label set now is actually the power set of the original label set, that's why we name this method as Label Powerset. A conventional multi-class classifier will be trained on this new dataset, and it will predict the possibilities of each possible subset of labels directly. Compared with BR method, the benefit of LP is that the label dependencies have been taken into consideration since it predicts multiple labels simultaneously. However, the number of labels would grow dramatically, and become huge especially when the m is large. So the instances with same label will be very few, and it's very difficult to build an effective multilabel classifier. In order to eliminate the disadvantages of BR and LP, various methods have been proposed consequently, some of which will be introduced in the following parts.

8.3.2 Classifier Chains and Probabilistic Classifier Chains

Classifier Chains (CC) algorithm was proposed by Read et al. recently based on BR method [13]. Similar with BR method, CC also generates a dataset and then trains a classifier for each label respectively, but it takes the dependencies among labels into consideration. CC methods mainly consists of the following steps. First of all, it randomizes all the labels and links them along a chain, and it assumes that each label is dependent on all its preceding labels in the chain. Secondly, the feature set of each label's corresponding dataset is extended with all its preceding labels, and each of these new features in every instance take value of 0/1, according to whether it is the instance's true label or not in original dataset. Finally, a binary classifier for label l_i is trained based on the new feature set consist of original features and all the labels l_i is dependent on, thus the label dependencies are incorporated in the process of training classifiers. Let us see an example of CC method using the dataset given in Section 8.1. We simply assume that the randomized order of labels is still $\{l_1, l_2, l_3, l_4\}$, thus the generated datasets for every label are given in Fig. 8.3.3.

When predicting labels for test instance x, these labels have to be predicted in the chain order. Therefore, predictions could be merged with x to form a new instance suitable for the next classifier. Although CC method could utilize label dependencies in a simple way, a potential problem is the labels are ranked randomly, so the learned dependencies might not always be consistent with the truth. Thus ECC (Ensemble of Classifier Chains) method was also proposed [13], which uses ensemble learning to learn multiple CC classifiers, each of which is trained using a different order

insts	l_1
x_1	1
x_2	1
x_3	0
x_4	0

(a) D_1

insts	l_1	l_2
x_1	1	0
x_2	1	1
x_3	0	1
x_4	0	1

(b) D_2

insts	l_1	l_2	l_3
x_1	1	0	1
x_2	1	1	0
x_3	0	1	0
x_4	0	1	0

(c) D_3

insts	l_1	l_2	l_3	l_4
x_1	1	0	1	0
x_2	1	1	0	1
x_3	0	1	0	0
x_4	0	1	0	1

(d) D_4

Figure 8.3.3: The transformation results using classifier chains

of labels. During the prediction phrase, results from all the classifiers will be averaged to eliminate the impact of randomness of labels' order and increase the overall accuracy.

For every label, CC method always selects the optimal value currently as its final prediction, without considering its influence on the following labels. So it's a greedy algorithm essentially, and might not always reach the globally optimal results. Inspired by CC method, Dembczynski et al. proposed the **probabilistic classifier chains** (PCC) method. It solves the multi-label problem from the viewpoint of risk minimization and Bayes optimal prediction [14]. PCC method uses Formula given in 8.3.1 to predict labels for test instances.

$$P(y \mid x) = \prod_{k=1}^{m} P(y_k \mid y_1, y_2, ..., y_{k-1}, x) \tag{8.3.1}$$

where $y_1, y_2, ..., y_{k-1}$ is label y_k's preceding labels in the chain. Similar with CC method, it also ranks the labels randomly, and assumes each label is dependent on all the preceding labels. The difference is that PCC will check all the possible label vector, that is 2^m paths in the search space, to find the one with the highest joint probability, whereas CC only follows a single path in the search space.

PCC method could achieve a better performance, but the possible improvement is at the cost of a much higher complexity approximate to $O(2^m)$. Hence it's reasonable to design more suitable methods that could depict the label dependencies in a lower-dimensional label space, where the computation cost is affordable.

8.3.3 Decompose the Label Set

We can view the aforementioned BR and LP methods as two extreme cases. While BR assumes labels are independent, LP assumes dependencies exist in any combination of labels. In most cases, the truth is that the labels could be divided into several groups. Labels from the same group are dependent on each other strongly, whereas labels from different groups are independent. Methods based on this assumption usually consist of two primary steps: one is how to partition the labels to determine the dependencies among labels, the other is how to incorporate such dependencies in the learning process. Based on such a learning framework, many approaches have been proposed by researchers. Let us look through several typical methods in this subsection.

RA*k*EL (RAndom *k* labELsets). This method was proposed based on the LP transformation [16]. As mentioned above, the number of labels may become very large and there would be not enough instances to train an effective classifier. Hence Tsoumakas et al. introduced the RA*k*EL method, aiming at reducing the computation cost of LP while keeping label dependencies will still be considered. The main idea in RA*k*EL is firstly to break a large set of labels into a number of small-sized label subsets randomly. The size of each of the subsets is k, a parameter that could be adjusted to reach the best performance. Then a LP classifier is trained for each of these label subsets. Let's see a example using the dataset given in 8.1. We simply assume that the set of labels $\{l_1, l_2, l_{3,4}\}$ are divided into two subsets: $\{l_1, l_3\}$ and $\{l_2, l_4\}$, so the generated datasets are as shown in Fig. 8.3.4.

insts	$\{l_1, l_3\}$
x_1	1
x_2	0
x_3	0
x_4	0

insts	$\{l_2, l_4\}$
x_1	0
x_2	1
x_3	0
x_4	1

(a) dataset for label $\{l_1, l_3\}$ (b) dataset for label $\{l_2, l_4\}$

Figure 8.3.4: The transformation results using RA*k*EL method

From Fig. 8.3.4, we could see that this method generates a dataset for each subset of labels, and the dataset contains all the instances but each instance's label is changed to 1 if all the labels in this group is its true labels, otherwise it is changed to 0. Then the LP method is applied to learn the labels in the same subset simultaneously. Thus the dependencies among labels within the same group are incorporated in the learning process, whereas the possible dependencies between labels from different groups are ignored. Moreover, each new label will be associated with multiple instances, since the size of label subset k, might be far less than m, the size of the original set of labels. For a test instance x, its final prediction is given by combining predictions of all the classifiers.

Labels in the above example are divided into two disjoint groups, however they can also be divided into overlapping groups. Since any label might appear in multiple overlapping groups, it will be predicted several times for a test instance. Thus the simple voting strategy will be used to determine whether it's true or false based on each value's times be predicted. Although having taken label dependencies into consideration, the primary problem of RAkEL method is that it just determines the division of labels randomly, so it might result in ignoring the true dependencies and generating the wrong dependencies instead.

Tenenboim et al. proposed a similar framework with RAkEL method [18]. In this framwork, Tenenboim assumes that label dependencies could be analysed explicitly in the following several ways: (1) Label distribution analysis; (2) Features among category distribution analysis; (3) Category combinations shown in the training set; (4) Supervised definition of dependencies. After determining the label dependency, one kind of clustering method could be used to cluster all the labels into disjoint subsets.

PS (Pruned sets) method has been proposed by Read et al. [15]. Similar with RAkEL, it's also based on the LP method and aims at solving the problem of infrequent label vectors and the difficulty to build effective classifiers. The difference is that RAkEL divides the labels into different subsets and generates a dataset for each subset, whereas PS method would prune the label sets until its occurrences $n > p$, a pruning threshold, and only one dataset is generated finally. The main process is described in the following steps:

(a) Let us $D = \{(x_1, C_1), (x_2, C_2), \ldots, (x_n, C_n)\}$ be a multi-label dataset. For every label vector C_i, its number of times of occurrences in the whole dataset D is computed and aggregated as n_i.

(b) For every label vector C_i, if its count $n_i > p$, then all the instances associated with it will be kept. Otherwise it will be decomposed into the subsets: $\{s_1, s_2, \ldots, s_n\}$, while any subset s_i's times of occurrences is greater than p.

(c) Furthermore, any instance x_i associated with c_i will be deleted from the dataset firstly, and then be copied and associated with each of these subsets, resulting in a set of instances $\{(x_i, s_1), (x_i, s_2), \ldots, (x_i, s_n)\}$. All these instances will be added to the dataset finally.

Now a single dataset is generated and each distinct label vector has been associated with enough instances, so an effective LP classifier could be trained based on it.

8.3.4 Transform Original Label Space to Another Space

Although we have used a simple dataset to explain a number of methods well, the number of labels in real problems may be huge, and it's usually very difficult to classify instances in a high-dimensional label space. Hence researchers try to transform the original label space into a new label space where the number of labels is smaller and the instances are more separable.

ECC framework (Error-Correcting Codes). Ferring et al. have applied the ECC framework on multi-label problems [19]. The main process of this method includes three steps:

(a) Firstly, an encoder $en(\bullet) : 0, 1^m \rightarrow 0, 1^k$ is used to transform each instance's original label vector y to a new label vector b, usually $b <<$ k. Thus each instance (x_i, y_i) is transformed to (x_i, b_i).

(b) Then, a multi-label classifier is learned based on the new dataset. In other words, instead of learning a classifier $f(x) : x \rightarrow y$, the objective is changed to learn a classifier $h(x) : x \rightarrow b$.

(c) During the prediction phase for a test, for instance x, we first predict the label vector b_x that might be associated with it, then use a decoder $den(\bullet) : 0, 1^k \rightarrow 0, 1^m$ to get the prediction in the original label space.

We can see that this is a general framework and any encoder, decoder and classifier can be used in it. It's also noted that when dividing labels into overlapped groups, the previous RA*k*EL method can also be seen as a special case of ECC method.

KDE-based methods (Kernel Density Estimation). Basic KDE method can be used to learn the dependencies between two classes of objects [20]. When used for multi-label classification, it's similar to the ECC framework and consists of the same steps. The difference is that the main object of KDE-based methods is to reduce the high dimensional label space into a low dimensional label space, to find the latent dependencies among labels and make the computation cost affordable. However, it's not necessary for the encoder used in ECC framework to reduce the label space's dimension, its main object is to make the instance more separable. The encoder used in KDE can also be implemented by various techniques. For instances, Hsu et al. used **CS (compressed sensing)** to perform a linear transformation on the label space [21]. Although the encoder of Compressive Sensing is linear, the decoder is not. It needs to solve an optimization problem when predicting labels, for instance, for something that is very time-consuming.

Motivated by the CS method, Tai et al. proposed the **PLST** (Principal Label Space Transformation) method [22], which also performs a linear transformation on the label space. It uses the principal components analysis technique as the encoder. In the viewpoint of linear algebra, we could create a $m \times n$ matrix Y based on a training dataset, where each column y_i is the label vector of the ith instance. Thereby, the problem of space transformation can be solved by finding an appropriate project matrix P to realize $H = P \bullet Y$, while H is a $k \times n$ matrix and $k \ll m$. Each column h_i of H is the ith instance's label vector in the low-dimensional label space. In PLST, the SVD decomposition is performed on the matrix Y, as shown in Formula 8.3.2.

$$Y = U\Sigma V^T \tag{8.3.2}$$

Here U is a $k \times k$ unitary matrix, and V is a $n \times n$ unitary matrix. The matrix is a k × n diagonal matrix that contains the singular values σ_i of each singular vector u_i in matrix U. Without loss of generality, we could assume that the singular values are ordered as $\sigma_1 \geq \sigma_2 \geq \ldots \geq \sigma_m$. Now Formula 8.3.2 can be rewritten as:

$$U^T Y = \Sigma V^T \tag{8.3.3}$$

Since the largest k singular values indicate the principle directions of the original label space Y, so we could discard the rest of the singular values and their corresponding singular vectors in U to get a smaller project matrix $P = U^T_k = [u_1, u_2, \ldots u_k]^T$ that projects the original label matrix Y to a new matrix $H = \Sigma V^T$. The decoder can also be obtained easily from Formula 8.3.3, it is P^{-1}, the inverse matrix of P. So we can see that it's a straightforward method, compared with CS method that has to solve an optimization problem.

8.4 Algorithm Adaptation

Algorithm adaptation is another fundamental strategy for multi-label classification. It extends conventional classification models, such as KNN, decision tree, Naive Bayes etc., to enable them to deal with the multi-label problems. Currently, a large number of single-label classification methods have been extended, let's see some representative examples in this section.

8.4.1 KNN-based methods

ML-KNN (Multi-Label K-nearest Neighbour) is the first multi-label lazy learning algorithm [10]. As its name implies, ML-KNN is derived from the popular K-nearest neighbor (KNN) algorithm. It decomposes a multi-label problem into multiple independent binary problems, each one corresponds to one label. ML-KNN firstly finds the K nearest neighbours in the training set for a test instance, then some statistics are collected and the principle of maximum a posteriori is used to determine the label set of the test instance.

Let x be an instance, y be the binary label vector associate with x, and $N(x)$ represents its k nearest neighbours in the training set. For each label l_i, ML-KNN will calculate the following statistics information first of all.

$$C_x(i) = \sum_{x \in N(x)} y_x(l_i) \tag{8.4.1}$$

where $y_x(l_i)$, the ith value of y_x, is 1 if label l_i is x's true label, and 0 else. Let H_x^i represent the event that l_i is x's true label, thus $P(H_x^i \mid C_x(i))$ represents the posteriori probability of H_x^i, given $C_x(i)$ instances in $N(x)$ are assigned with label l_i, and $P(\neg H_x^i \mid C_x(i))$ represents the posteriori probability of H_x^i is false. Thereby, for label l_i, we could get the classification function f_i for it:

$$f_i(x) = \begin{cases} 1 & \text{if } P(H_x^i \mid C_x(i)) > P(\neg H_x^i \mid C_x(i)) \\ 0 & \text{else} \end{cases}$$

Classification is simple right now. For an instance x and a label l_i, we firstly calculate $C_x(i)$, $P(H_x^i \mid C_x(i))$ and $P(\neg H_x^i \mid C_x(i))$, then l_i is predicted as the x's true label if $P(H_x^i \mid C_x(i)) > P(\neg H_x^i \mid C_x(i))$, otherwise as a false label. The key issue is how to calculate these probabilities. Using Bayesian rule, $P(H_x^i \mid C_x(i))$ can be rewritten as

$$P(H_x^i \mid C_x(i)) = \frac{P(H_x^i)\, P(C_x(i) \mid H_x^i)}{P(C_x(i))} \tag{8.4.2}$$

where $P(H_x^i)$ is event H_x^i's prior probability, and $P(C_x(i) \mid H_x^i)$ is the probability that the number of instances in $N(x)$ which are associated with label y_i is $C_x(i)$, given H_x^i is true. We can estimate the $P(H_x^i)$ and $P(C_x(i) \mid H_x^i)$ by counting the frequency in training set. Specifically speaking, the $P(H_x^i)$ could be estimated by

$$P(H_x^i) = \frac{s + \sum_{i=1}^{n} y_x(l_i)}{s \times 2 + n} \tag{8.4.3}$$

Here s is a smoothing parameter and is usually set to 1, that is the Laplace smoothing. Estimation of $P(C_x(i) \mid H_x^i)$ is a little bit complicated. For label $l_i (1 \leq i \leq m)$, there will be an array: k_i, which has $k+1$ elements. The value of k_i's jth element is the count of occurrences that label l_i is a true label of current instance, as well as j instances from its k nearest neighbouring. Then we could get the conditional probability $P(C_x(i) \mid H_x^i)$ as Equation 8.4.4.

$$P(C_x(i) \mid H_x^i) = \frac{s + k_j[C_j]}{s \times (k+1) + \sum_{r=0}^{k} k_j[r]} \tag{8.4.4}$$

Similarly, we could compute $P(\neg H_x^i \mid C_x(i))$ and $P(C_x(i) \mid \neg H_x^i)$ in the same way. Since $P(C_x(i))$ is constant, so we can simply compare $P(H_x^i) \times P(C_x(i) \mid H_x^i)$ and $P(\neg H_x^i) \times P(C_x(i) \mid \neg H_x^i)$ to get the prediction.

We can see that ML-KNN method learns a classifier for each label, so actually it can also be viewed as a problem transformation method. Although it has been recognized as an effective algorithm, it assumes the labels are independent of each other and could not utilize the dependencies among labels to facilitate the learning process.

Similar with ML-KNN, Cheng et al. proposed **IBLR-ML**, a method that is also based on the kNN method [23]. The difference is that it takes label dependencies into consideration and combines model-based and similarity-based inference for multi-label classification. IBLR-ML firstly uses the stacking framework to give a base prediction for each label, and then gives the final prediction for each label based on the base predictions of other labels. Thus the prediction process for a test instance x consists of two steps. Firstly, basic KNN method is used to predict base labels of x respectively. For instance, for label $l \in \{0, 1\}$, its probability to be x's base label is given by the Equation 8.4.5.

$$P(l = 1) = \frac{|\{x_i \in N(x) \mid y_l = 1\}|}{N(x)} \tag{8.4.5}$$

Here $N(x)$ is x's k nearest neighbours. According to KNN model, the probability of label l to be x's base label is the proportion of instances in $N(x)$ that are also associated with l.

IBLR-ML then uses a logistic regression model to compute the final probability of each label l_i based on the base probabilities of all labels. Let bp_i and fp_i denote the base and final probability respectively, that l_i is x's true label. The final probability can then be got by the Equation 8.4.6.

$$\log(\frac{fp_i}{1 - fp_i}) = a_0 + \sum_{i=1}^{m} a_i bp_i \tag{8.4.6}$$

Here $a_0, \ldots, a_m$ is the parameters needed to be learned. Now it is clear that when giving final prediction for each label, all other labels are considered, thus the potential dependencies are considered.

8.4.2 Learn the Label Dependencies by the Statistical Models

So far, most of aforementioned methods assume that labels are independent or learn the dependencies in a intuitive way. Formal definition of label dependency is not given and how to measure and depict it is also not clear. Thus researches applied a number of statistical models to depict the label dependencies explicitly.

Tenenboim-Chekina et al. proposed the ChiDep algorithm. It measures the dependencies between pairwise labels using Chi-square test and divides the labels into several mutually exclusive subsets [24]. Ghamrawi et al. applied the conditional random fields model to create an undirected graphical representation of the relationships between labels and features [25]. Bieza et al. proposed the multi-dimensional Bayesian network [26], which organizes the labels and features into three subgraphs: label subgraph, feature subgraph, and label-feature subgraph. Zhang et al. proposed the LAED method [27], which uses a Bayesian network to represent the relationships between labels. Fu et al. proposed the LDTS method to depict the label dependencies using a tree model [28]. Guo et al. used the conditional dependency network to create a cyclic directed graphic model for representing the label dependencies [29].

8.5 Evaluation Metrics and Datasets

Evaluation is a critical way to find the appropriate methods for a specific application. Now we have encountered different kinds of multi-label learning methods and there are more potential methods we haven't mentioned. With lots of available methods, we have to determine which one could generate the most appropriate classification model for a application in practice. You may wish to evaluate and compare their performances to find the good ones. But what is good and how we can estimate it? Moreover, in most cases we could characterize multi-label instances from different aspects and these features are related to the model's performance closely. So what are the representative features of multi-label data and how could we get them? These questions will be addressed in this section. First, some typical evaluation metrics are introduced, then multiple datasets that are widely used for evaluating different multilabel learning methods and several statistics for characterizing these datasets are described.

8.5.1 Evaluation Metrics

Since each instance may have several labels simultaneously, conventional metrics used for evaluating single-label classification methods, such as accuracy, error rate, etc. can not be used directly. Thus several appropriate metrics that can deal with multi-label predictions are introduced by researchers. Generally speaking, these metrics mainly fall into two categories, i.e., bipartition-based metrics and ranking-based metrics. Furthermore, they can be further divided as instance-based metrics and label-based metrics [9]. Before going through the detail definitions of these metrics, we first specify the notations that will be used. Let $D = \{(x_1, C_1),$

$(x_2, C_2), \ldots, (x_n, C_n)\}$ be a dataset, where x_i is its ith instance, and $C_i \subset L$ is its true labels. Given a classifier f and a test instance x_i, Y_i denotes the possible labels of x_i predicted by f. $rank(x_i)$ or $rank_i$ denotes the predicted rank of labels, and $rank(x_i, l)$ denotes the label l's position in the rank.

Bipartition-based metrics simply focus on whether the labels are correctly predicted or not, without considering to what extent the labels are correctly predicted. Thus predictions with different confidences will get the same evaluation value according to these metrics. Several well-known metrics of this type are as follows:

(1) *Subset accuracy*. It is similar to the *accuracy* used in single-label learning, and computes the ratio of instances for which the predicted set of labels match the true set of labels exactly. Its definition is given by the Equation 8.5.1.

$$SubsetAccuracy(f, D) = \frac{1}{n} \sum_{i=1}^{n} I(Y_i = C_i) \tag{8.5.1}$$

where $I(true)=0$ and $I(false)=0$. Obviously it's a very strict measure, since the prediction will still be viewed as totally wrong, even if only few of the labels are predicted incorrectly and the rest get right predictions. The greater this measure is, the better the classifier's performance is, and the optimal value could be 1, that indicates all instances' labels get the exact predictions. However, it' should be noted that it's very difficult to predict all labels correctly when there are a huge number of labels, thus the actual value should be very small in most cases.

(2) *Hamming Loss*. This measure is proposed by Schapire and Singer [3], and it's definition is given in Equation 8.5.2.

$$H\text{-}Loss(f, D) = \frac{1}{n} \sum_{i=1}^{n} \frac{|Y_i \oplus C_i|}{m} \tag{8.5.2}$$

where the operator $\oplus$ calculates the symmetric difference of two sets of labels. So $| \bullet |$ returns the number of misclassified labels, for instance. We can see that it's a more reasonable measure compared with *Subset accuracy*, since it results in better evaluation to the classifier which can predict the majority of the labels correctly. The smaller this measure is, the better the classifier's performance, and the optimal value could be 0 when the labels of all instances are predicted correctly. *Hamming Loss* can also be viewed as a kind of label-based metric, since it can be decomposed by labels, and for each label the evaluation is the same as the traditional *accuracy* for single-label learning.

(3) *Precision*. It calculates the ratio between intersection of the two sets of labels and the set of true labels, as depicted in Equation 8.5.3.

$$Precision(f, D) = \frac{1}{n}\sum_{i=1}^{n}\frac{|Y_i \cap C_i|}{|C_i|} \tag{8.5.3}$$

As shown, this metric would compute the portion of one instance's true sets of labels that are predicted correctly. The difference between it and *Hamming Loss* is it only concerns the predictions of one instance's set of true labels and does not care about whether the remaining labels are predicted correctly or not, while *Hamming Loss* takes all the labels' predictions into consideration. The greater this measure is, the better the classifier's performance is, and the optimal value could be 1 when all the instances' true labels get right predictions.

(4) *Recall*. Different from *Precision*, this metric calculates the ratio between intersection of the two sets of labels and the set of predicted labels, as showed in Equation 8.5.4.

$$Recall(f, D) = \frac{1}{n}\sum_{i=1}^{n}\frac{|Y_i \cap C_i|}{|Y_i|} \tag{8.5.4}$$

It evaluates the classifier's performance from the perspective of predicted labels, since it only focuses on the portion of an instance's set of predicted labels that are its true labels. The greater this measure is, the better the classifier's performance, and the optimal value could be 1 when all the predicted labels are the instance's true labels, even some of its true labels are still predicted wrongly.

(5) F_1 measure. Since *precision* and *recall* evaluate classifier's from different prospectives, optimizing any one will make the other decline. Therefore, F_1 measure is introduced to make a trade-off between them and get a reasonable result. It's described by Equation 8.5.5.

$$F_1(f, D) = \frac{1}{n}\sum_{i=1}^{n}\frac{2|Y_i \cap C_i|}{|Y_i|+|C_i|} \tag{8.5.5}$$

the better the classifier's performance is, and the optimal value could be 1. The aforementioned 3 metrics are heavily used in information retrieval to evaluate the returned documents given an ad hoc query.

(6) *Accuracy*. It evaluates the average ratio of the intersection of the two sets of labels and the union of the two sets of labels, as depicted in Formula 8.5.6.

$$Accuracy(f, D) = \frac{1}{n}\sum_{i=1}^{n}\frac{2|Y_i \cap C_i|}{|Y_i|+|C_i|} \tag{8.5.6}$$

The greater this measure is, the better the classifier's performance, and the optimal value could be 1. We can see that it's a more strict metric compared with *precision* and *recall*, since only when the predicted label set matches the true label set exactly, the optimal value could be reached.

All the above are bipartition-based metrics. There are other kinds of metrics called rank-based metrics. While the former are based on binary predictions of the labels, the latter are based on a predicted rank of labels, instead of giving an explicit *yes/no* predictions. The following are the several commonly used rank-based metrics.

(1) *One-error.* It evaluates how many times the top-ranked label is not in the instance's set of true labels, as given in Equation 8.5.7.

$$One\text{-}error(f, D) = \frac{1}{n} \sum_{i=1}^{n} \delta(\arg \max_{l \in Y_i} rank(x_i, l)) \qquad (8.5.7)$$

where $\delta(\bullet) = 1$ when l isn't x_i's true label, otherwise $\delta(\bullet) = 0$. *One-error* is not a very rigorous metric, since it only concerns whether the top-rank one is a true label or not, while ignoring the remaining true labels' predictions. So it might not able to give a reasonable evaluation of the classifier's performance. The smaller this measure is, the better the classifier's performance, and the optimal value could be 0 when the top-rank label is the its true label for all instances.

(2) *Coverage.* It computes how far it is needed to go down the ranked list of labels to cover one instance's all true labels, as given in Formula 8.5.8

$$Coverage(f, D) = \frac{1}{n} \sum_{i=1}^{n} \max_{l \in C_i} rank(x_i, l) - 1 \qquad (8.5.8)$$

Smaller value of this measure means more true labels are ranked before the false labels and thus the classifier's performance is better.

(3) *Ranking Loss.* It computes the number of times when false labels are ranked before the true labels, as given in Equation 8.5.9.

$$R\text{-}Loss = \frac{1}{n} \sum_{i=1}^{n} \frac{1}{|C_i \| \overline{C_i}|} |\{(l_a, l_b) : rank(x_i, l_a) > rank(x_i, l_b), (l_a, l_b) \in C_i \times \overline{C_i}\}| \qquad (8.5.9)$$

where $\overline{C_i}$ is the set of false labels of instance x_i. This metric compares any possible pairwise labels' rank that one is the true label and the other is a false label. The smaller this measure, the better the classifier's performance, and the optimal value could be 0 when any one of the true labels is ranked before all the false labels.

(4) *Average Precision.* This metric is firstly used in the field of information retrieval, to evaluate the rank of result documents, given a specific query. It computes the average fraction of labels ranked above a

particular true label, while these labels are also true labels. The definition is given in Formula 8.5.10.

$$\text{AvePrec} = \frac{1}{n}\sum_{i=1}^{n}\frac{1}{|C_i|}\sum_{l\in C_i}\frac{|\{l' \in C_i : rank(x_i,l') \le rank(x_i,l)\}|}{rank(x_i,l)} \tag{8.5.10}$$

The greater this measure, the better the classifier's performance is, and the optimal value could be 1.

Roughly speaking, all preceding bipartition-based and rank-based metrics evaluate the performance of classifiers from different aspects. Therefore, there could be no single classification model which could perform well on all metrics, and there is also no general metric that could be used for evaluating any kinds of classifiers. It depends on the particular problems and the objectives of learning to select appropriate metrics. Moreover, there might be potential relationships between different metrics, which are implicit now and need further investigation.

8.5.2 Benchmark Datasets and the Statistics

In this subsection, some representative datasets used extensively by researchers are presented. Since different characteristics of multi-label data might have different impacts on the learning methods, so the explicit definitions of various characteristics should be given first to assist observation of the relationship between datasets and classifiers' performances. The following are a number of fundamental statistics for characterizing multi-label datasets.

(1) *Label Cardinality*. It calculates the average number of labels for each instance, which is defined as follows.

$$\text{LC} = \frac{1}{n}\sum_{i=1}^{n}|C_i| \tag{8.5.11}$$

where $|C_i|$ is the number of the ith instance's true labels.

(2) *Label Density*. It is calculated through dividing the label cardinality by m, the size of original label set, which is defined as follows.

$$\text{LD} = \frac{1}{n}\sum_{i=1}^{n}\left|\frac{C_i}{m}\right| \tag{8.5.12}$$

(3) *Distinct Label Sets*. It counts the number of distinct label vectors which appeared in the data set, which is defined as follows:

$$\text{DLS}(D) = |\{C\,|\,\exists(x, C) \in D\}| \tag{8.5.13}$$

(4) *Proportion of Distinct Label Sets.* It normalizes the $DLS(s)$ by the number of instances, which is defined as follows.

$$\text{PDLS}(D) = \frac{DLS*(S)}{n} \tag{8.5.14}$$

Table 8.1: Description of representative multi-label benchmark datasets

name	insts	atts	labels	LC	LD	DLS	PDLS
bibtex[7]	7395	1836	159	2.402	0.015	2856	0.386
bookmarks[7]	87856	2150	208	2.028	0.010	18716	0.213
CAL500[31]	502	68	174	26.044	0.150	502	1
corel5k[32]	5000	499	374	3.522	0.009	3175	0.635
delicious[5]	16105	500	983	19.020	0.019	15806	0.981
emotions[33]	593	72	6	1.869	0.311	27	0.045
enron	1702	1001	53	3.378	0.064	753	0.442
genbase[34]	662	1186	27	1.252	0.046	32	0.516
mediamill[35]	43907	120	101	4.376	0.043	6555	0.149
medical	978	1449	45	1.245	0.028	94	0.096
rcv1v2[36]	6000	47236	101	2.880	0.029	1028	0.171
tmc2007[37]	28596	49060	22	2.158	0.098	1341	0.046
scene[12]	2407	294	6	1.074	0.179	15	0.006
yeast[38]	2417	103	14	4.237	0.303	198	0.081

As mentioned in the beginning, multi-label data are ubiquitous in the real applications, including text analysis, image classification, prediction of gene functions etc. Researchers have extracted multiple benchmark datasets from these practical problems and used them to examine and compare various multi-label methods' performance. Tsoumakas et al. have summarized a number of datasets used commonly, with corresponding informations including source reference, number of instances, features, labels, etc. and other statistics [9]. Table 8.1 gives the detailed descriptions of the datasets and all of them are available for download at the homepage of Mulan, an open source platform for multi-label learning [30].

8.6 Chapter Summary

So far we have introduced the definition of multi-label classification and various types of algorithms for it. The typical measure metrics and datasets used for experiments are also given. Although we have reached a great deal of achievements, the problem is actually not tackled very well. The following issues still should be given enough concerns and need further research.

The first one is the instance spareness problem, since the number of possible label vectors has grown explosively with the increasing size of original label space m. For example, the size of possible label space would be 2^{20} finally, even if m is only 20. Consequently, the number of positive instances that have the same label vector would decrease dramatically, and it is difficult to build an effective classifier for each possible label vector.

The second one is the label spareness problem. In most cases, there will be hundreds or even thousands of labels in the label space, whereas most of the instances might only be associated with few labels only, for example, less than five. Hence most of the labels will have few positive instances and too many negative instances. This situation would lead to the class-imbalance problem that is common in machine learning when building classifiers label by label, and make learning the real distribution of labels more difficult.

Moreover, there are always certain kinds of dependencies among the labels in multi-label data. For instance, a thriller is likely to be an action film, while the same book could probably not be of technique and fiction simultaneously. It can be seen clearly that learning these dependencies would benefit the learning process. Although many methods have been proposed, many of them simply assume that there are only random label dependencies, not giving a formal definition of dependency and measure it precisely. Thus the dependencies these method have learned might violate the reality. Other primary challenges and practical issues include the curse of dimension, how to explore the semantic meaning of labels etc., thus more appropriate classification models for multi-label problems are needed.

References

[1] J. Han and M. Kamber. *Data mining concepts and techniques*. San Mateo, CA: Morgan Kaufmann. 2006.

[2] A. K. McCallum. Multi-label text classification with a mixture model trained by EM. *In: Proceedings of AAAI'99 Workshop on Text Learning*. 1999.

[3] R. E. Schapire and Y. Singer. Boostexter. A boosting-based system for text categorization. Machine Learning, Vol. 39, pp. 135–168, Number 2–3, 2000.

[4] A. Clare and R. D. King. Knowledge discovery in multi-label phenotype data. *In: Proceedings of the 5th European Conference on Priciples of Data Mining and Knowledge Discovery(PKDD2001)*. Freiburg, Germany, pp. 42–53, 2001.

[5] G. Tsoumakas, I. Katakis and I. Vlahavas. Effective and efficient multilabel classication in domains with large number of labels. *In: Proceedings of ECML/PKDD 2008 Workshop on Mining Multidimensional Data (MMD08)*, pp. 30–44, 2008.

[6] G. Qi, X. Hua, Y. Rui et al. Correlative multi-label video annotation. *In: Proceedings of the 15th International Conference on Multimedia*. New York, pp. 17–26, 2007.

[7] I. Katakis, G. Tsoumakas and I. Vlahavas. Multilabel text classification for automated tag suggestion. *In: Proceedings of the ECML/PKDD 2008 Discovery Challenge*. 2008.

[8] Y. Song, L. Zhang and C. L. Giles. Automatic tag recommendation algorithms for social recommender systems. *ACM Transactions On the Web*, Vol. 1, pp. 4, Number 5, 2011.

[9] G. Tsoumakas, I. Katakis and I. Vlahavas. Mining multi-label data. *In:* M. Oded and R. Lior (eds.). *Data Mining and Knowledge Discovery Handbook*, New York: Springer, pp. 667–85, 2010.

[10] M. Zhang and Z. Zhou. ML-KNN: A lazy learning approach to multi-label learning. Pattern Recognition, Vol. 40, pp. 2038–2048, Number 7, 2007.

[11] M. Zhang, J. M. Pena and V. Robles. Feature selection for multi-label naive bayes classification. *Information Sciences*, Vol. 179, pp. 3218–29, Number 19, 2009.

[12] M. R. Boutell, J. Luo and X. Shen et al. Learning multi-label scene classification. *Pattern Recognition*, Vol. 37, pp. 1757–71, No. 9, 2004.

[13] J. Read, B. Pfahringer and G. Holmes et al. Classifier chains for multi-label classification. In: *Proceedings of the European Conference on Machine Learning and Knowledge Discovery in Databases*: Part II. Bled, Slovenia: Springer-Verlag, pp. 254–69, 2009.

[14] K. Dembczynski, W. Cheng and E. Hullermeier. Bayes optimal multilabel classification via probabilistic classifier chains. In: J. Furnkranz and T. Joachims (eds.). *Proceedings of the 27th International Conference on Machine Learning (ICML2010)*. Aifa, Israel: Omnipress, pp. 279–86, 2010.

[15] J. Read. Multi-label classification using ensembles of pruned sets. In: *Proceedings of the 2008th IEEE International Conference on Data Mining(ICDM2008)*. Pisa, Italy, pp. 995–1000, 2008.

[16] G. Tsoumakas, I. Katakis and I. Vlahavas. Random k-labelsets for multi-label classification. IEEE Transactions on Knowledge and Data Engineering, Vol. 23, pp. 1079–89, Number 7, 2011.

[17] W. Chen, J. Yan and B. Zhang. Document transformation for multi-label feature selection in text categorization. In: *Proceedings of the 7th IEEE International Conference on Data Mining*. Omaha, NE, pp. 451–56, 2007.

[18] L. Tenenboim, L. Rokach and B. Shapira. Multi-label classification by analyzing labels dependencies. In: *Proceedings of the Workshop on Learning from Multi-label Data at ECML PKDD 2009*. Bled, Slovenia, pp. 117–131, 2009.

[19] C. Ferng and H. Lin. Multi-label classification with error-correcting codes. *Journal of Machine Learning Research*, Vol. 20, pp. 281–95, 2011.

[20] J. Weston, O. Chapelle and A. Elisseeff. Kernel dependency estimation. In: S. Becker, S. Thrun and K. Obermayer (eds.). Advances in Neural Information Processing Systems 15 (NIPS 2002), pp. 873–80, 2003.

[21] D. Hsu, S. Kakade and J. Langford. Multi-label rrediction via compressed sensing. In: Y. Bengio, D. Schuurmans and J. D. Lafferty (eds.). Proceedings of 24th Annual Conference on Neural Information Processing Systems. Vancouver, British Columbia, Canada, pp. 772–80, 2009.

[22] F. Tai and H. Lin. Multi-label classification with principle label space transformation. In: *2nd International Workshop on Learning from Multi-Label Data (MLD10)*. Haifa, Israel, pp. 45–52, 2010.

[23] W. Cheng and E. Hullermeier. Combining instance-based learning and logistic regression for multilabel classification. *Machine Learning*, Vol. 76, pp. 211–25, Number 2–3, 2009.

[24] L. Tenenboim-Chekina, L. Rokach and B. Shapira. Identification of label dependencies for multi-label classification. In: 2nd International Workshop on Learning from Multi-label Data (MLD'10). Haifa, Isrel, pp. 53–60, 2010.

[25] N. Ghamrawi and A. K. McCallum. Collective multi-label classification. In: *Proceedings of the 2005 ACM Conference on Information and Knowledge Management (CIKM2005)*. Bremen, Germany, pp. 195–200, 2005.

[26] C. Bielza, G. Li and P. Larranage. Multi-dimensional classification with Bayesian networks. *International Journal of Approximate Reasoning*, Vol. 52, pp. 705–727, No. 6, 2011.

[27] M. Zhang and K. Zhang. Multi-label learning by exploiting label dependency. In: B. Rao, B. Krishnapuram, A. Tomkins et al. (eds.). Proceedings of the 16th ACM SIGKDD International Conference on Knowledge Discovery and Data Mining (KDD2010). Washington, DC, USA, pp. 999–1008, 2010.

[28] Bin Fu, Zhihai Wang, Rong Pan et al. Learning tree structure of labels dependency for multi-label learning. In: P. Tan, S. Chawla and C. K. Ho (eds.). *Proceedings of 16 Pacific-Asia Knowledge Discovery and Data Mining (PAKDD 2012)*. Kuala Lumpur, Malaysia, pp. 159–170, 2012.

[29] Y. Guo and S. Gu. Multi-label classification using conditional dependency networks. *In:* T. Walsh (eds.). *Proceedings of the 22nd International Joint Conference on Artificial Intelligence(IJCAI2011).* Barcelona, Catalonia, Spain, pp. 1300–05, 2011.

[30] G. Tsoumakas, E. Spyromitros-Xioufis and J. Vilcek Mulan. A java library for multilabel learning. *Journal of Machine Learning Research,* Vol. 12, pp. 2411–14, 2011.

[31] D. Turnbull, L. Barrington and D. Torres. Semantic annotation and retrieval of music and sound effects. IEEE Transactions On Audio, Speech, and Language Processing, Vol. 16, pp. 467–76, Number 2, 2008.

[32] P. Duygulu, K. Barnard and J. F. G. de Freitas. Object recognition as machine translation: Learning a lexicon for a fixed image vocabulary. *In:* A. Heyden, G. Sparr and M. Nielsen (eds.). *Proceedings of the 7th European Conference on Computer Vision.* Copenhagen, Denmark, pp. 97–112, 2002.

[33] K. Trohidis, G. Tsoumakas, G. Kalliris et al. Multi-label classication of music into emotions. *In:* J. P. Bello, E. Chew and D. Turnbull (eds.). *Proceedings of the 9th International Conference on Music Information Retrieval (ISMIR2008).* Philadelphia, PA, USA, pp. 325–30, 2008.

[34] S. Diplaris, G. Tsoumakas and P. Mitkas. Protein classification with multiple algorithms. *In: Proceedings of the 10th Panhellenic Conference on Informatics (PCI 2005).* Volos, Greece, pp. 448–56, 2005.

[35] C. G. M. Snoek, M. Worring and J. C. van Gemert. The challenge problem for automated detection of 101 semantic concepts in multimedia. *In: Proceedings of the 14th annual ACM international conference on Multimedia,* New York, USA, pp. 421–30, 2006.

[36] D. D. Lewis, Y. Yang and T. G. Rose. RCV1: a new benchmark collection for text categorization research. *Journal of Machine Learning Research,* Vol. 5, pp. 361–97, 2004.

[37] A. Srivastava and B. Zane-Ulman. Discovering recurring anomalies in text reports regarding complex space systems. *In: Proceedings of the 2005 IEEE Conference on Aerospace.* pp. 3853–62, 2005.

[38] A. Elisseeff and Jason Weston. A kernel method for multi-labelled classification. *In:* T. G. Dietterich, S. Becker and Z. Ghahramani (eds.). *Advances in Neural Information Processing Systems 14.* Vancouver, British Columbia, Canada, MIT Press, pp. 681– 87, 2001.

CHAPTER 9

Privacy Preserving in Data Mining

Privacy preserving in data mining is an important issue because there is an increasing requirement to store personal data for users. The issue has been thoroughly studied in several areas such as the database community, the cryptography community, and the statistical disclosure control community. In this chapter, we present the basic concepts and main strategies for the privacy-preserving data mining.

The k-anonymity approach will be presented in Section 9.1. The l-diversity strategy will be introduced in Section 9.2. The t-Closeness method will be presented in Section 9.3. Discussion on privacy preserving data mining will be presented in Section 11.4. Chapter summary will be presented in Section 11.5.

9.1 The K-Anonymity Method

Due to the importance of privacy preserving in various applications, especially for protecting personal information, Samarati [21] first introduced the issue and proposed several efficient strategies to address it. Samarati observed that although the data records in many applications are made public by removing some key identifiers such as the name and social-security numbers, it is not difficult to identify the records with the help of taking into account some other public data. This happens especially more commonly in the medical and financial field where microdata that are increasingly being published for circulation or research, can lead to abuse, compromising personal privacy.

Figure 9.1.1 shows a concrete example to explain the personal information leaking issue. The published data in Fig. 9.1.1 (a) has been de-identified by removing the users' names and Social Security Numbers (SSNs). It is thus thought to be safe enough. Nevertheless, some attributes of the public data, such as ZIP, DateOfBirth, Race, and Sex can also exist in other public datasets and therefore, this information can be jointly used to identify the concrete person. As shown in Fig. 9.1.1 (a), the attributes of ZIP, DateOfBirth, and Sex in the Medical dataset can be linked to that of the Voter List (Fig. 9.1.1 (b)) to discover the corresponding persons' Name, Party, and so forth. Given the concrete example, we can observe that there is one female in the Medical dataset who was born on 09/15/61 and lives in the 94142 area. This information can uniquely recognize the corresponding record in the Voter list, that the person's name is Sue J. Carlson and her address is 900 Market Street, SF. Through this example, we can see that there exists personal information leak by jointly considering the public data.

SSN	Name	Race	DateOfBirth	Sex	Zip	Martial Status	HealthProblem
		asian	04/18/64	male	94139	married	chest pain
		asian	09/30/64	female	94139	divorced	obesity
		black	03/18/63	male	94138	married	shortness of breath
		black	09/07/64	female	94141	married	obesity
		white	05/14/61	male	94138	single	chest pain
		white	09/15/61	female	94142	widow	shortness of breath
		....	....	....	....	....	....

(a) Medical dataset

Name	Address	City	ZIP	DOB	Sex	Party	....
....	....	....	....	....	....	....	....
....	....	....	....	....	....	....	....
Sue J. Carlson	900 Market St.	San Francisco	94142	09/15/61	female	democrat	....
....	....	....	....	....	....	....	....

(b) Voter list

Figure 9.1.1: Re-identifying anonymous data by linking to external data [21]

To address the problem mentioned above, Samarati [21] introduced an effective concept, i.e., k-anonymity, defined as follows:

Definition 1 *(k-anonymity) Each release of the data must be such that every combination of values of quasi-identifiers can be indistinguishably matched to at least k respondents.*

The key idea is that, to reduce the risk of record identification, it requires that each record in the public table should be not distinct and no fewer than k records can be returned according to any query.

To fulfill the purpose of k-anonymity, the author in [21] introduced two main strategies, i.e., generalization and suppression. For the generalization approach, it generalizes the attribute values of records to a larger range so that the granularity of the representation is reduced. For the above example, it could generalize the date of birth to the year of birth and therefore, make those records indistinct. The idea of the suppression strategy is to remove those sensitive attributes' value or hide them. As we can see, the privacy obtained from these strategies is at the price of losing some information of the original data. Therefore, there is always a trade-off between privacy preserving and the accuracy of the transformed data. To take a good balance between privacy and accuracy, the author in [21] introduced the concept of k-minimal generalization to limit the level of generalization while keeping as much data information as possible with regard to some determined anonymity.

Although the work in [21] has tackled the privacy preserving issue for some extent, the problem itself is very difficult to solve optimally. It is well known that the problem of optimal k-anonymization is NP-hard, as demonstrated in [20]. As a result, most existing studies aim to introduce effective and efficient heuristic strategies to address it, such as [21, 5, 13]. More detail survey about the issue can be found in [6].

Bayardo et al. [5] introduced an order-based strategy to improve the efficiency of tackling the issue. It takes the attributes of records into two groups, i.e., quantitative attribute and categorical attribute. For the quantitative attributes, the values of them are discretized into intervals. For the categorical attributes, the values are clustered into different classes. The authors deal with each group as an item that could be ordered. Similar to traditional database techniques, the authors [5] introduced an effective index to facilitate the traversing process on a set enumeration tree. This tree is in a similar sense of that applied in the frequent pattern mining literature (See Chapter 6), which is used to enumerate all the candidate generalizations based on the items. The construction of the tree is as follows: (1) first set the root of the tree, which is a null node; and (2) each successive level of the tree is built by adding one item which is larger than all the items in the previous tree. The order is based on the lexicographical order. We can see that it could be possible the tree grows too large and thus, could be impractical to deal with. To address this issue, the authors [5] proposed several effective strategies to prune the candidate generalizations as early as possible. However, all of these techniques are heuristic and the complexity of the tree building (and item generalization) is not optimal. The proposed strategy in [5] follows a branch and bound manner, that it could terminate the item generalization process earlier. As demonstrated in the paper [5], the introduced algorithm shows a good performance compared with the state-of-the-art ones.

To further improve the efficiency, in a later paper [13], LeFevre et al. proposed the *Incognito* algorithm. The basic idea of *Incognito* is that it utilizes bottom-up breadth first search strategy to traverse all the candidate generalizations. Specifically, it generates all minimal k-anonymous tables through the following steps: (1) for each attribute, it removes those generalizations which could not satisfy the k-anonymity; (2) it joins two (k)-dimensional generalizations to obtain the (k+1)-dimensional candidate generalization and then evaluate the candidate. If it cannot pass the k-anonymity test, the candidate will be pruned. This step is in a similar sense of that introduced for frequent pattern mining (i.e., candidate-generate-and -test in Chapter 6). All the candidate generalizations can be traversed without duplication and loss. There is a distinct for [14, 13] that the authors deal with the data as a graph instead of a tree, which was assumed in the previous work [21].

There are some other works applying the generalization and suppression strategies to tackle the privacy preserving issue [23, 10]. The basic idea of [10] is that it applies a top down approach to traverse all the candidate generalization. Because of the special property of the process, which is to reverse the bottom up process, it will decrease the privacy and increase the accuracy of the data while traversing the candidates. As stated in the paper [10], the method can control the process so as to obey the k-anonymity rule. In a later paper [23], the authors introduced several complementary strategies, e.g., bottom-up generalization, to further improve the whole performance.

The essential step to generate the candidate generalization is to traverse all the subspaces of multi-dimensions. As a result, it could use genetic algorithm or simulated annealing to tackle the issue. Iyengar [11] introduced the genetic algorithm based strategy to transform the original into k-anonymity model. In another work [24], the authors proposed a simulated annealing algorithm to address the problem.

In addition to the commonly used strategies, i.e., generalization and regression, there is some other techniques proposed, such as the cluster based approach [8, 2, 3]. The basic idea for these works is that the records are first clustered and each cluster is represented by some representative value (e.g., average value). With the help of these pseudo data, privacy can be effectively preserved while the aggregation characteristics of the original data is well reserved. However, how to measure the trade-off between the privacy and the reserved data information seems to be an issue.

	Non-Sensitive			Sensitive
	Zip Code	Age	Nationality	Condition
1	13053	28	Russian	Heart Disease
2	13068	29	American	Heart Disease
3	13068	21	Japanese	Viral Infection
4	13053	23	American	Viral Infection
5	14853	50	Indian	Cancer
6	14853	55	Russian	Heart Disease
7	14850	47	American	Viral Infection
8	14850	49	American	Viral Infection
9	13053	31	American	Cancer
10	13053	37	Indian	Cancer
11	13068	36	Japanese	Cancer
12	13068	35	American	Cancer

(a) Inpatient Microdata

	Non-Sensitive			Sensitive
	Zip Code	Age	Nationality	Condition
1	130**	<30	*	Heart Disease
2	130**	<30	*	Heart Disease
3	130**	<30	*	Viral Infection
4	130**	<30	*	Viral Infection
5	1485*	≥40	*	Cancer
6	1485*	≥40	*	Heart Disease
7	1485*	≥40	*	Viral Infection
8	1485*	≥40	*	Viral Infection
9	130**	3*	*	Cancer
10	130**	3*	*	Cancer
11	130**	3*	*	Cancer
12	130**	3*	*	Cancer

(b) 4-anonymous Inpatient Microdata

Figure 9.2.1: Example for l-diversity [17]

Using views appropriately is another technique to protect privacy. The basic idea is that we can just show a small part of the views (that sensitive attributes can be controlled) to the public. However, this approach may fail if we unintentionally publish some important part of the views, which lead to the violation of k-anonymity. [26] studied the issue of using multiple views and clarified that the problem is NP-hard. Moreover, the authors introduced a polynomial time approach if the assumption of existing dependencies between views holds.

The complexity of tackling the k-anonymity issue is difficult to measure. The existing works limit the analysis on the approximation algorithms [4, 3, 20]. These methods guaranteed the solution complexity to be within a certain extent. In [20], the authors introduced an approximate method that at $O(k \cdot logk)$ cost, while in [4, 3], the authors proposed some algorithms that guaranteed $O(k)$ computation cost.

9.2 The l-Diversity Method

Although the k-anonymity is simple and effective to tackle the issue of privacy preserving to some extent, it is susceptible to many vicious attacks, such as homogeneity attack and background knowledge attack [17, 18], defined as follows.

- Homogeneity Attack: For this case, there are k tuples that have the same value of a sensitive attribute. From the previous viewpoint, it follows the privacy preserving of the k-anonymity. However, these k tuples as a group can be identified uniquely.
- Background Knowledge Attack: For this case, it is possible that the sensitive and some quasi-identifier attributes can be combined together to infer certain values of some sensitive attributes.

The concrete examples which describe the attacks are illustrated in Fig. 9.2.1. It shows the patient information from a New York hospital (Fig. 9.2.1 (a)). There are no critical attributes such as name, SSN, and so forth. The attributes are classified into two categories, i.e., the sensitive and non-sensitive attributes. The values of the sensitive attributes are preferred by adversaries. Figure 9.2.1 (b) presents the 4-anonymity transformed data, i.e., the mark * indicates a suppression value between 0 and 9 for Zip code and Age. The examples for the attacks which cannot be prevented by the k-anonymity are shown as follows:

- Example of Homogeneity Attack [17]: Alice and Bob are neighbors who know each other very well. Alice finds the 4-anonymous table, i.e., Fig. 9.2.1 (b), which is published by the hospital and she knows that Bob's information exists in it. Moreover, Alice knows that Bob is an American whose age is 31 and lives in the 13053 area. As a result, it is easy to infer that Bob's number is between 9 and 12. From the table, Alice can make a conclusion that Bob has cancer because any person whose number in the range (i.e., [9,12]) has the same health problem.

- Example of Background Knowledge Attack [17]: Alice has another friend, Tanaka, whose medical information also appears in the table (i.e., Fig. 9.2.1 (b)). Tanaka is a Japanese female whose age is 21 and lives in the 13068 area so it can be inferred that Tanaka's number is between 1 and 4, whose health problem could be heart disease or viral infection. Because it is well known that Japanese people seldom got heart disease, Alice knows that Tanaka has a viral infection issue.

From the above examples, we can see that although the k-anonymity is an effective solution to preserve the records' privacy, in some cases, it may lose effectiveness on protecting sensitive information. To tackle these issues, l-diversity [17, 18] was proposed by keeping the diversity of the attributes to hide sensitive information.

Machanavajjhala et al. [17, 18] proposed the diversity principle, i.e., l-diversity, to protect sensitive information from malicious attacks. To fulfill this purpose, it requires each quasi-identifier attribute group has at least l "well represented" different values, which can be used to make the tuple indistinct. To define how well the values of attributes represented, several possible models could be applied. The simplest one is that at least l distinct values exist in the attribute group. If we have $l=k$, then it should satisfy the k-anonymity. This model is mentioned in [22]. However, this simple implementation can still be attacked, i.e., probabilistic inference attack. The reason is that some values are more frequent than others in the group and it is not difficult to deduce those frequent ones based on the distribution of the values. To tackle this issue, some more deliberated principles based on l-diversity are introduced.

- Entropy l-diversity [17]: For entropy l-diversity, in each quasi-identifier group, we have $-\sum_{s \in S} P(qid, s) log(P(qid, s)) \geq log(l)$, where S is a sensitive attribute, and $P(qid, s)$ is the probability of tuples in a quasi-identifier group which have the value s. Because of the property of entropy, we can know that a larger value of the entropy indicates the sensitive values can distributed more evenly in the group, which makes the tuples more indistinct. However, the entropy l-diversity is still not perfect to prevent all the attacks. Moreover, it has the drawback that the entropy value is difficult to understand for users, who prefer some probability based explanation, i.e., malicious attackers have 20% chance to know that Bob has cancer, according to the current l-diversity setting.
- Recursive (c, l)-diversity [17]: Similar to the entropy l-diversity, the key idea of the recursive (c, l)-diversity also ensures that the sensitive values are distributed as more evenly as possible, that the frequent values are not so frequent and the rare values are not so rare. In a given quasi-identifier group qn, r_i is denoted as the number of times the ith

most frequent sensitive value appears in qn. Given a constant c, qn satisfies recursive (c, l)-diversity if $r_1 < c(r_l + r_{l+1} + \cdots + r_m)$. A table satisfies the recursive (c, l)-diversity if every quasi-identifier group satisfies the recursive -diversity. Note that 1-diversity is always satisfied.

There are some other principles based on the l-diversity introduced, i.e., positive disclosure-recursive (c, l)-diversity and negative/positive disclosure-recursive (c_1, c_2, l)-diversity [17, 18]. The basic idea of these principles is that although with the help of some background an attacker may remove some values from the group, he still cannot recognize sensitive information. From another viewpoint of these principles, some work is introduced to estimate the maximum disclosure risk of the published data [19] based on different privacy preserving metrics.

Note that the above mentioned issues are based on such an assumption that only one sensitive attribute exists. If there are multiple sensitive attributes, the l-diversity problem becomes more challenging. Some works have explored this issue, i.e., [18, 25]. However, all these works suffer from the issue of the curse of dimensionality.

9.3 The *t*-Closeness Method

Due to the intrinsic drawback of the l-diversity, Li et al. [16] found that leakage of sensitive information could happen when the overall distribution of a sensitive attribute is skewed. The reason is that because the l-diversity requirement ensures "diversity" of sensitive values in each group, it does not take into account the semantical closeness of these values. For example, suppose we have a patient table where 90% of the tuples have headache and 10% have cancer. If we have a quasi-identifier group which has 50% of headache and 50% of cancer, it satisfies the 2-diversity rule. Nevertheless, this quasi-identifier group may face to a privacy problem because it is easy to infer that any person in this group has 50% chance to get cancer, yet consider in the whole table, this probability reduces to 10%. The obvious difference between these two conclusions makes the l-diversity principle lose its effect.

To protect the attack from the above mentioned example, Li et al. [16] introduced the *t*-closeness principle. The key idea is that *t*-closeness requires the distribution of each sensitive attribute in every group should be similar to that in the overall table. Moreover, in [16], the authors introduced a new distance metric, i.e., Earth Mover Distance (EMD), to estimate the closeness between two distributions. A constant *t* is used as a threshold to satisfy the *t*-closeness principle. Although the advantage it may obtain, there are several issues brought by this interesting principle: (1) it is not easy to protect privacy according to different security levels; (2) the introduced distance metric, i.e., EMD, lacks a flexibility to cope with numerical attributes [15]; and (3) the utility of the published data may largely be sacrificed because it is too strict a rule to let all the distributions of attributes be similar to each other. Several solutions are introduced to tackle part (if not all) of these issues [9].

9.4 Discussion and Challenges

In this chapter, we presented the main strategies for privacy preserving data mining. There are several issues which should be mentioned. The first one is how to keep a good balance between different evaluation metrics, such as privacy and utility. It is intuitive that the safest strategy to protect privacy is to publish as few data as possible, though this approach, leads to low utility. For some principle (e.g., entropy *l*-diversity), how to explain the semantic meaning of the setting becomes more difficult and thus, is challenging to be applied on the real applications. Another main issue for privacy preserving is the curse of dimensionality. The work in [1] states that to keep privacy, a large number of the attributes may need to be suppressed or generalized. This requirement also leads to the loss of the data's utility. More seriously, it seems that some methods become infeasible to implement with more dimensionality taken into account.

9.5 Chapter Summary

In this chapter, we introduced the basic concept and main techniques for the privacy-preserving data mining issue. We presented a variety of data transformation strategies such as *k*-anonymity, *l*-diversity, and *t*-closeness based methods. Furthermore, we gave some concrete examples to illustrate the advantages and disadvantages of these approaches and analyzed them thoroughly. Some related issues, i.e., curse of dimensionality and balance between utility and privacy, were also discussed.

References

[1] C. C. Aggarwal. On k-anonymity and the curse of dimensionality. *In: Proceedings of the 31st International Conference on Very Large Data Bases*, pp. 901–909, 2005.

[2] C. C. Aggarwal and P. S. Yu. A condensation approach to privacy preserving data mining. *In: Proceedings of International Conference on Extending Database Technology*, pp. 183–199, 2004.

[3] G. Aggarwal, T. Feder, K. Kenthapadi, S. Khuller, R. Panigrahy, D. Thomas and A. Zhu. Achieving anonymity via clustering. *In: Proceedings of the Twenty-fifth ACM SIGMOD-SIGACT-SIGART Symposium on Principles of Database Systems*, pp. 153–162, 2006.

[4] G. Aggarwal, T. Feder, K. Kenthapadi, R. Motwani, R. Panigrahy, D. Thomas and A. Zhu. Anonymizing tables. *In: T. Eiter and L. Libkin, editors, ICDT*, pp. 246–258, 2005.

[5] R. J. Bayardo and R. Agrawal. Data privacy through optimal k-anonymization. *In: Proceedings of the 21st International Conference on Data Engineering*, pp. 217–228, 2005.

[6] V. Ciriani, S. D. C. di Vimercati, S. Foresti and P. Samarati. *Anonymity*, Vol. 33 of *Advances in Information Security*. Springer, 2007.

[7] C. Clifton. Using sample size to limit exposure to data mining. *J. Comput. Secur.*, 8: 281–307, December 2000.

[8] J. Domingo-Ferrer and J. M. Mateo-Sanz. Practical data-oriented microaggregation for statistical disclosure control. *IEEE Trans. on Knowl. and Data Eng.*, 14: 189–201, January 2002.

[9] J. Domingo-Ferrer and V. Torra. A critique of k-anonymity and some of its enhancements. *In: Proceedings of the 2008 Third International Conference on Availability, Reliability and Security*, pp. 990–993, 2008.

[10] B. C. M. Fung, K. Wang and P. S. Yu. Top-down specialization for information and privacy preservation. *In: Proceedings of the 21st International Conference on Data Engineering*, pp. 205–216, 2005.

[11] V. S. Iyengar. Transforming data to satisfy privacy constraints. *In: Proceedings of the eighth ACM SIGKDD International Conference on Knowledge Discovery and Data Mining*, pp. 279–288, 2002.

[12] L. V. S. Lakshmanan, R. T. Ng and G. Ramesh. To do or not to do: the dilemma of disclosing anonymized data. *In: Proceedings of the 2005 ACM SIGMOD International Conference on Management of Data*, pp. 61–72, 2005.

[13] K. LeFevre, D. J. DeWitt and R. Ramakrishnan. Incognito: efficient full-domain kanonymity. *In: Proceedings of the 2005 ACM SIGMOD international conference on Management of data*, pp. 49–60, 2005.

[14] K. LeFevre, D. J. DeWitt and R. Ramakrishnan. Mondrian multidimensional k-anonymity. *In: Proceedings of the 22nd International Conference on Data Engineering*, 2006.

[15] J. Li, Y. Tao and X. Xiao. Preservation of proximity privacy in publishing numerical sensitive data. *In: Proceedings of the 2008 ACM SIGMOD International Conference on Management of Data*, pp. 473–486, 2008.

[16] N. Li, T. Li and S. Venkatasubramanian. t-closeness: Privacy beyond k-anonymity and l-diversity. *In: Proceedings of the 21st International Conference on Data Engineering*, pp. 106–115, 2007.

[17] A. Machanavajjhala, D. Kifer, J. Gehrke and M. Venkitasubramaniam. ?-diversity: Privacy beyond k-anonymity. *In: Proceedings of International Conference on Data Engineering*, 2006.

[18] A. Machanavajjhala, D. Kifer, J. Gehrke and M. Venkitasubramaniam. L-diversity: Privacy beyond k-anonymity. *ACM Trans. Knowl. Discov. Data*, 1, March 2007.

[19] D. J. Martin, D. Kifer, A. Machanavajjhala, J. Gehrke and J. Y. Halpern. Worst-case background knowledge in privacy. *In: Proceedings of International Conference on Data Engineering*, pp. 126–135, 2007.

[20] A. Meyerson and R. Williams. On the complexity of optimal k-anonymity. *In: Proceedings of the twenty-third ACM SIGMOD-SIGACT-SIGART Symposium on Principles of Database Systems*, pp. 223–228, 2004.

[21] P. Samarati. Protecting respondents' identities in microdata release. *IEEE Trans. on Knowl. and Data Eng.*, 13: 1010–1027, November 2001.

[22] T. M. Truta and B. Vinay. Privacy protection: p-sensitive k-anonymity property. *In: Proceedings of the 22nd International Conference on Data Engineering Workshops*, 2006.

[23] K. Wang, P. S. Yu and S. Chakraborty. Bottom-up generalization: A data mining solution to privacy protection. *In: Proceedings of the Fourth IEEE International Conference on Data Mining*, pages 249–256, 2004.

[24] W. E. Winkler. Using simulated annealing for k-anonymity, 2002.

[25] X. Xiao and Y. Tao. Anatomy: simple and effective privacy preservation. *In: Proceedings of the 32nd International Conference on Very Large Data Bases*, pp. 139–150, 2006.

[26] C. Yao, X. S. Wang and S. Jajodia. Checking for k-anonymity violation by views. *In: Proceedings of the 31st international conference on Very large data bases*, pp. 910–921, 2005.

Part III

Emerging Applications

CHAPTER 10

Data Stream

Data stream mining is an important issue because it is the basis for numerous applications, such as network traffic, web searches, sensor network processing, and so on. Data stream mining aims to determine the patterns or structures of continuous data. Such patterns of structures may be used later to infer possible events that could occur. Data streams exhibit unique dynamics in that such data can be read only once. This feature presents a limitation to numerous traditional strategies from analyzing data streams because such techniques always assume that all data could be stored in limited storage. Thus, data stream mining could be considered as the performance of computations on a large amount of data or even unlimited data. In this chapter, we will introduce the basic concepts and main strategies that can be employed to address the aforementioned challenge.

The general data streaming models will be introduced in Section 10.1. The sampling approach will be presented in Section 10.2. The wavelet method will be discussed in Section 10.3. The sketch method will be presented in Section 10.4. The histogram method will be introduced in Section 10.5. A discussion on data stream will be presented in Section 11.4. A chapter summary will be presented in Section 11.5.

10.1 General Data Stream Models

Several models have been introduced in the data stream literature [53]. Given an input stream, S, the items arrive sequentially, that is, $a_1, a_2, \ldots, a_n$. Each item, a_i, describes a corresponding underlying signal A_i. Different models are distinct in terms of describing signals based on the items in the stream.

- **Time Series Model.** Each item a_i is the same as signal A_i. The items are received based on an increasing order of i. Numerous applications such as stock price stream and Web server log fit this kind of model well.

- **Cash Register Model [35].** In this model, the items a_i are accumulated to the signals A_j. Similar to a cash register, multiple a_i could be aggregated and set to a given A_j over time. This model can be applied to applications such as counting the access 198 number of the same IP address to a website.
- **Turnstile Model [53].** In this model, the items a_i are updates to A_j. The update operator can be insertion or deletion. The model is the most generalizable in that it fits a large number of applications such as the dynamic people situation in a subway system.
- **Sliding Window Model.** In this model, the mapping or computation of items is focused over a fixed-sized window in the stream. While the stream is in progress, items at the end of the sliding window are deleted, and the new items from the stream are considered. This kind of model fits applications such as weather prediction, which require the most up-to-date data stream.

As introduced in [53], the models in decreasing order of generality are as follows: turnstile, cash register, and time series[1]. Designing appropriate algorithms specific models is more practical, and the challenge lies in making these approaches sufficiently generalizable for strong models such as the turnstile.

10.2 Sampling Approach

Sampling is an important approach that is employed for numerous applications, such as signal processing, information survey, computer graphics, and so on. Sampling is based on the assumption that directly dealing with an extremely large amount of data is impractical, and therefore, some form of approximation is necessary. In statistics, sampling is concerned with the selection of a subset of items from a large data set to retain (and further measure) the properties of the whole data [4].

Sampling has the following advantages: (1) low cost; (2) efficient data storage; and (3) convenience in addressing the sampling data because of the small size. Compared with all other commonly used techniques for data streams, such as wavelets, sketches, histograms, the sampling strategy is probably the easiest and the most applicable approach, especially for challenging issues such as high dimensional data.

From a statistical perspective, the sampling technique aims to store the posterior distribution of the data stream to retain similar expectations and variances. Therefore, the expectation and variance of some function $f(\alpha)$ with

---

[1]The sliding window model can be considered as the constrained versions of the above three models.

respect to a probability distribution $p(\alpha)$, where α denotes the components that could be continuous or discrete, must be determined. To measure the expectation, we have the following formula [12]:

$$E[f] = \int f(\alpha)p(\alpha)d\alpha \qquad (10.2.1)$$

where the integral is the summation if α represents discrete variables. After sampling, we need to ensure that the new expectation of the sampled items is stored the same manner as that of the whole data set (i.e., $E[\hat{f}]=E[f]$). The new expectation estimator (i.e., $\hat{f}$) based on the samples α^n (where n ranges from 1 to N) relative to discrete variables is defined as follows:

$$\hat{f} = \frac{1}{N}\sum_{n=1}^{N} f(\alpha^n) \qquad (10.2.2)$$

Further, the variance of the new expectation estimator must also be the same as the former value (i.e., $var[\hat{f}]=var[f]$) and can be represented as follows:

$$var[\hat{f}] = \frac{1}{N}E[(\hat{f}-E[\hat{f}])^2] \qquad (10.2.3)$$

Thus, different sampling strategies have been introduced with a focus on how to select sample instances to satisfy the given rules, although these rules may vary for different applications and domains (e.g., sum aggregation). Another challenging issue is that the variables considered are not always independent, thus resulting in more complex estimation.

For a few applications, in addition to the expectation $E[f]$ and the variance $var[f]$, we need to measure the mean μ (i.e., $\mu=E[f]$) and the standard deviation σ (i.e., $\sigma = \sqrt{var[f]}$), which are commonly used to describe the distribution of the total data set. We introduce a number of inequalities, such as the Markov and Chebychev inequalities, to estimate these features [8].

To estimate the bound of the random variable α, we can use the Markov inequality:

$$P(\alpha > \beta) \leq \mu/\beta \qquad (10.2.4)$$

where β is a random variable of the data stream. The Chebychev inequality can be obtained by employing the Markov inequality for the random variable $(\alpha - \mu)^2/\sigma^2$:

$$P(|\alpha - \mu| > \beta) \leq \sigma^2/\beta^2 \qquad (10.2.5)$$

The Markov and Chebychev inequalities have been proven to be sufficiently generalizable. Specific applications such as Chernoff bound and Hoeffding inequality may be more suited to tighter bounds. For these bounds, we can employ the Markov inequality based on the parameterized functions relative to the specific applications. Moreover, the manner by which to select a sample from the stream is another issue, that is, the size of the sample must be determined. Intuitively, this size is determined by the total size of the stream, which cannot be known apriori. Therefore, the probability that any item is stored as a sample should be dynamically changed according to the data stream.

10.2.1 Random Sampling

From a statistics perspective, random sampling denotes the selection of a group of individuals (samples) randomly from a large data set (i.e., data stream). The probabilities for all individuals to be chosen during the process should be the same [67, 65]. Given its simplicity, random sampling can be a basis or component of other more sophisticated sampling methods.

A concrete example to illustrate the idea of random sampling is as follows [1]. Suppose that n students wish to obtain tickets for a football game. However, the tickets (i.e., M) are insufficient. Therefore, the students have to employ a fair method to determine the persons who can go. To fulfill this goal, every student is randomly given a number (i.e., from 0 to $n-1$). The students who obtain the first (or the last) M numbers are the winners.

In numerous applications, this kind of sampling is commonly applied *without replacement*, i.e., any number will not be chosen more than once. Under this assumption, the probability of one instance being chosen by the process is no longer independent, but the result is still reasonable, especially in the case of selecting a small group of samples from a large data set, because the probability of choosing the same individual is low. However, for other applications, sampling with a replacement strategy may be more appropriate.

The given example is built upon the assumption that the data are static, that is, the size of the data set is known apriori. However, for a data stream, the selection of individual samples is more challenging because the process is performed under a dynamically changing environment [8].

Suppose we want to extract M individual samples from a data stream. While receiving the data, we maintain a list of size M to store the candidate samples. In the initialization stage, the first M individuals in the data stream are stored in the list. As more data are received, we need to determine whether the next individual should be stored. The probabilities of storing these individuals as samples should be the same (i.e., M/n, where n is the size of the individuals received thus far in the data stream). Considering the replacement model in the process, an old individual has to be removed from the list before a new individual is stored as a sample. As proven by Aggarwal [8], the probability of storing any individual as sample in the list is $M/(n+1)$.

As introduced in Section 10.1, the assumed model is a time series, in which the individual item is received with the same importance. However, for a few applications, recent individual items may be of importance. We can use a sliding window to partition the data stream, and more recent windows are given larger weight relative to the results of queries. This issue has been explored by a number of researchers [32, 13, 11]. The sliding window-based model is more practical for real data stream applications considering the limitation resources such as the main memory and CPU.

10.2.2 Cluster Sampling

Cluster sampling [3] is another commonly used technique in the literature that was introduced mainly for static data [48]. The total data set is first clustered intuitively into several groups, and the representative individual of each group can then be deemed as a sample. The clusters should be mutually exclusive and collectively exhaustive. This method is cost effective, and the criteria for clustering on the data is domain specific, including time, position, nationality, and so on.

Considering that several clustering algorithms (e.g., [9]) take into account the dynamic property of data, the clustering sampling strategy can be implemented on data streams.

Compared with random sampling, cluster sampling generally needs more samples to achieve the same effectiveness (i.e., accuracy) because these samples are necessary to distinguish clusters from one another. Moreover, the implementation of cluster sampling is always conducted through multi-steps: the first step aims to build the clusters that will subsequently be used; in the second step, primary individuals are randomly selected as samples for each group; and in the following steps, we recursively determine whether other individuals from the selected clusters are samples. This kind of multi-step sampling can largely reduce sampling cost.

Another sampling technique called *stratified sampling* [5], is similar to cluster sampling. This approach first partitions the whole data into homogeneous subgroups before sampling using two criteria: (1) the strata should be mutually exclusive, such that every individual in the data set must be assigned to only one stratum; and (2) the strata should be collectively exhaustive, such that no individual can be excluded. These criteria are similar to those of cluster sampling. Finally, random sampling can be used for each stratum. Through these strategies, stratified sampling can improve the representativeness of the samples by reducing sampling error.

Cluster and stratified sampling have a number of key differences: (1) in cluster sampling, the cluster is treated as the sampling unit, and the analysis is executed on the level of clusters, whereas in stratified sampling, the analysis is implemented on the individuals in the strata; (2) in stratified sampling, a sample is randomly chosen from each strata, whereas in cluster sampling, only the randomly selected clusters are explored; and (3) cluster sampling primarily aims to reduce costs by increasing sampling efficiency, whereas stratified sampling aims to increase effectiveness (i.e., precision). However, both of these sampling methods are limited by the unknown size of the total data.

As introduced in [27], several issues confront existing sampling techniques. First, data streams have an unknown dataset size. Therefore, the sampling process on a data stream requires a special analysis to limit the error bounds. Another problem is that to check the sampling strategy may be inappropriate for checking anomalies in surveillance analysis because the data rates in the stream are always changing. Thus, we explore the relationship among the data rate, sampling rate, and error bounds for real applications.

10.3 Wavelet Method

The wavelet-based technique is a fundamental tool for analyzing data streams. From a traditional perspective, a wavelet is a mathematical function used to divide a given function or continuous time series into different scale components [6]. This approach has been successfully applied to applications such as signal processing, motion recognition, image compression, and so on [63, 7]. The wavelet technique provides concise and general summarization of data (i.e., stream), which can be used as the basis for efficient and accurate query processing methods. Numerous strategies have been introduced based on the idea of wavelet, in which the most commonly used approach for data streams is called *Haar wavelets* [68].

The Haar wavelet provides a foundation for query processing on stream and relational data. It creates a decomposition of the data (or compact summary) into a set of Haar wavelet functions, which can be used for later query processing. The essential step is the determination of the Haar wavelet coefficients. Only coefficients with high values are typically stored. Higher order coefficients in the decomposition generally indicate broad trends in the data, whereas lower order coefficients represent the local trends. We will show a concrete example to illustrate the Haar wavelet process [66, 31].

Suppose our data stream is {3, 2, 4, 3, 1, 5, 0, 3}. The data in the vector are computed as averaged values between neighbors to obtain a lower resolution representation (i.e., level 2) of the data, such as $\left[\frac{3+2}{2}, \frac{4+3}{2}, \frac{1+5}{2}, \frac{0+3}{2}\right] = \left[\frac{5}{2}, \frac{7}{2}, 3, \frac{3}{2}\right]$. This transformation results in the loss of information, thus requiring more information to be stored. The Haar wavelet technique computes the differences of the averaged values between neighbors, such as $\left[\frac{3-2}{2}, \frac{4-3}{2}, \frac{1-5}{2}, \frac{0-3}{2}\right] = \left[\frac{1}{2}, \frac{1}{2}, -2, -\frac{3}{2}\right]$. These eight data values are the first-order coefficients of the sample data, which can be used to recover the original data set. Similarly, we can obtain the lower resolution representation (i.e., formal part of level 1) as $\left[\frac{\frac{5}{2}+\frac{7}{2}}{2}, \frac{3+\frac{3}{2}}{2}, \frac{\frac{5}{2}-\frac{7}{2}}{2}, \frac{3-\frac{3}{2}}{2}\right] = \left[3, \frac{9}{4}, -\frac{1}{2}, \frac{3}{4}\right]$.

Recursively, we can obtain the final Haar wavelet transformation of the data as $\left\{\frac{21}{8}, \frac{3}{8}, -\frac{1}{2}, \frac{3}{4}, \frac{1}{2}, \frac{1}{2}, -2, -\frac{3}{2}\right\}$. The whole transformation process is illustrated in Fig. 10.3.1.

Original data (level 3)	3	2	4	3	1	5	0	3
level 2	2.5	3.5	3	1.5	0.5	0.5	-2	-1.5
level 1	3	2.25	-0.5	0.75				
level 0	2.625	0.375						

Haar wavelet transformation on {3, 2, 4, 3, 1, 5, 0, 3}

Figure 10.3.1: Sample Haar wavelet transformation

Haar wavelet analysis commonly assumes that the size q of the time series data is a power of two without loss of generality because the series can be decomposed into segment subseries, each of which has a length that is a power of two. From the process of Haar wavelet transformation, we obtain 2^{l+1} coefficients of level l (as shown in Fig. 10.3.1). Each coefficient (i.e., 2^{l+1}) represents (and summarizes) a contiguous part of the data stream (i.e, $q/2^{l+1}$). In the segment series data, the ith of the 2^{l+1} coefficients covers the part beginning from $(l + 1) \cdot q/2^{l+1}+1$ to $i \cdot q/2^{l+1}$.

Previous works were not concerned about retaining all coefficients, but only a smaller number of them (i.e., top-B) [36]. Given this simplicity, some information on the original data during the transformation will be lost. In [36], the authors report that the highest B-term approximation is in fact the best B-term approximation that it minimizes the sum squared error for a given B.

A large number of existing approaches employ a lossy mechanism because of the large number of coefficients introduced by the Haar wavelet transformation, that is, the number is equal to the length of the total data stream. While keeping the top-B coefficients with large values, the other ones are set to zero. This heuristic is thus, employed to reduce the dimensionality of the time series data. However, a trade-off always exists between the number of coefficients and the error introduced by the transformation. Obtaining the optimal number of Haar wavelet coefficients is an interesting issue in the literature. Nevertheless, previous works assume that only a small number of coefficients dominate the full effectiveness of the transformation.

In addition to the determination of the optimal number of the Haar wavelet coefficients, another issue is the selection of the appropriate coefficients. The absolute value is not the sole optimal criterion to dominate performance. For example, based on different evaluation metrics for various applications such as mean square error vs. least maximum error, different selection strategies may be appropriate [29, 30, 63]. However, other issues such as computation efficiency should also be considered.

As introduced in [8], other important topics are related to Haar wavelet transformation. A number of applications prefer to monitor large quantities of information simultaneously in the same stream data source. Taking sensor applications, the monitors may store a large number of data features, such as location, pressure, wind direction, and so on. Therefore, the Haar wavelet transformation needs to process all of these features simultaneously. An intuitive approach is the application of decomposition on each feature, such that the top-B coefficients are discovered by merging the transformed results [63]. However, this strategy may be inefficient, because duplicate transformations may be conducted on the same individual data (relative to different features). The authors in [22] introduced several effective strategies for the simultaneous monitoring and transformation of multi-feature data by using bitmaps to determine the optimal features.

In summary, Haar wavelet transformation continuously builds a summarization of the B representative wavelet coefficients (e.g., with largest absolute values) for time series data set. Considering the special properties of dynamic stream data, the following criteria should be satisfied: (1) sub-linear space usage must be available to store the summarization and (2) sub-linear per-item update time must be sufficient to maintain the summarization. Applications on query processing may further need to consider another criterion, that is, sub-linear query time.

10.4 Sketch Method

The sketch-based technique is one of the major tools for stream data analysis. Sketches are small space summarizations of stream data in a centralized or distributed environment. The main advantage of sketch-based techniques is that they require storage that is significantly smaller than the input stream length. For most sketch based algorithms, the storage usage is sub-linear in N, that is, $log^k N$, where N is the input size and k is some constant. The hashing function is generally employed by sketch based algorithms to project the data stream into a small space sketch vector that can be easily updated and queried. As demonstrated by numerous experimental evaluations, the cost of updating and querying on the sketch vector is only a constant time for each operation.

Given the property of a sketch, the answers to the queries that are determined by examining the sketches are only approximations because only part of the data information is stored (as sketch). A sketch generally contains multiple counters for random variables relative to different attributes. The error boundary on the answers should be held with probabilistic guarantees. The introduced sketch-based algorithms differ in terms of defining and updating random variables as well as in the efficient querying of the sketches.

The sketch-based technique is closely related to the random projection strategy [45]. Indyk et al. [41] introduced this strategy into the database domain (i.e., time series domain) to discover the representative trends. The key idea is that a data point with dimensionality d is reduced by (randomly) selecting k dimensionalities. The dot product of the these k dimensionalities between data points is computed. Each k dimensionality (i.e., random vectors) follows the normal distribution with zero mean and unit variance. Moreover, the random vector is normalized relative to one unit in magnitude. Accuracy is dependent on the value of k, where a larger value of k results in high accuracy. We will then introduce several sketch-based algorithms for processing data streams. Please refer to [17, 62, 8] for a more detailed survey of related issues.

10.4.1 Sliding Window-based Sketch

Indyk et al. [41] introduced the sketch-based technique into the database domain to discover the trends governing data streams. The authors observed that the length of a time series can be considered to have one dimensionality. Therefore, we can construct the sketch by considering the length as the random vector. Two situations are considered in [41]: fixed window sketches and variable window sketches.

For fixed window sketches, the aim is to obtain sliding window sketches with a fixed length l. Thus, $l \cdot k$ operations should be conducted for a sketch with size k. Given the total of $O(n-l)$ sliding windows, $O(n \cdot l \cdot k)$ operations are necessary. From this analysis, we find that if the window length l is large (i.e., the same order of magnitude as the time series), the cost of calculation would be quadratic to the size of the series. Therefore, this approach could be impractical for large time series data (i.e., stream). As introduced in [41], the construction of fixed window-based sketches can be considered as the computation of the polynomial convolution of random vectors of appropriate length over the time series data. Therefore, we can use the fast Fourier transform to address the issue. This observation indicates that the fixed window sketches can be obtained efficiently.

For variable window sketches, the aim is to construct the sketches for any sub-vector between length l and u [41]. This approach requires $O(n^2)$

sub-vectors that may have the length $O(n)$ in the worst case. Through this analysis, we find that the total cost is $O(n^3)$, which limits the application of the technique for large time series data (i.e., stream). To address this issue, Indyk et al. [41] proposed the construction of a set of sketches. The size of the group must be considerably smaller than the total sketches. To determine the sketches to be stored, the authors deliberately chose sub vectors out of the original ones, such that the computation can be achieved in $O(1)$ time with sufficient accuracy guaranteed for each vector. The experimental evaluation demonstrates the effectiveness and efficiency of the introduced strategies [41]. Please refer to [41] for more details about this technique.

10.4.2 Count Sketch

Alon et al. [10] were the first to introduce the term sketch as a tug-of-war sketch. The authors aim to measure and optimize the second order of the frequency moment $F_2 = \sum_i f_i^2$. Notably, more recent studies found that the introduced summarization in [10] can also be utilized to measure the inner product of two distributions on frequency, that is, $\sum_i f_i f_i'$, where f_i and f_i' denote the two frequency distributions. From the observation, we find that if f_i can be obtained from a data stream, the product of $f_i'=1$ and $f_j'=0$ for all $j \neq i$ can be calculated during the query processing. Thus, we find that for the error bound, the value should be $\epsilon F^{1/2}{}_2 \leq \epsilon n$, which has more than $1-\delta$ probability for a sketch with size of $O(\frac{1}{\epsilon^2} \log 1/\delta)$.

The cost of updating the count sketch is high because all the sketches should be rebuilt if a new instance comes in. Thus, the naive count sketch technique as an appropriate strategy for data streams. To address the issue, Charikar et al. [14] proposed an improved algorithm that requires only a small part of the sketches to be updated when a new instance comes in. Thus, performance is significantly improved.

We briefly describe the idea in [14]. Interested readers are referred to the previous paper for more details. The introduced sketch structure contains a $d \times \omega$ array (denoted as C) that stores counters (where d is the number of rows), with two hash functions presented for each row. One hash function g maps the instances of the data stream into $[\omega]$, whereas the other hash function h maps the instances into $\{-1, +1\}$. For each row j (i.e., $1 \leq j \leq d$), the corresponding mapping on instance i, that is, $h_j(i)$, is stored in the array element $C[j, g_j(i)]$. We find that $\widehat{f}_i$ is median $1 \leq j \leq d$ $h_j(i)C[j, g_j(i)]$ [17]. The analysis shows that for each value of j, the expectation and variance depending on F_2/ω can be accurately derived.

10.4.3 Fast Count Sketch

To improve the efficiency of count sketches further, Thorup and Zhang [64] introduced the fast count sketch technique, that uses one random hashing to hasten the update time while maintaining reasonable error bounds. The price of obtaining this improvement is that more sketch vectors are utilized, and deliberately designing the hash function is necessary. In the fast count sketches, the counters in the vector are the same as those of count sketches, the only difference is that the former contains a four-universal hash function that is associated with the vector.

When a new data instance i arrives, its mapped value, ω, is immediately stored into the corresponding counter, that is, $x_t[h(i)] = x_t[h(i)] + \omega$, where $h : I \rightarrow \{1, \ldots, n\}$ is the four-universal hash function [64].

We can deduce the estimate of the size of the join attributes based on the second frequency moment as in [62]. As claimed by the authors, the estimate is an unbiased one of the inner product $\bar{f} \circ \bar{g}$. The variance is retained in the fast count sketches in a manner similar to that in the count sketches. The multiplicative factor is $\frac{1}{n-1}$ for the fast count sketches but $\frac{1}{n}$ for the count sketches. More entries are needed in the fast count sketches than in the count sketches, but the difference is negligible for large values of n.

10.4.4 Count Min Sketch

Cormode and Muthukrishnan [18] introduced another effective sketch type, that is, count min sketches, for facilitating the construction and update of the synopsis. The data structure of count min sketches is the same as that of fast count sketches. Count min sketches apply a series of two-universal functions to map the data instances, which differs from the four-universal functions used in fast count sketches. The mechanism of sketch update in Count Min sketches is the same as that in fast count sketches.

An issue for count min sketches is that they employ the L_1 norm, whereas count sketches utilizes the L_2 norm. Thus, count min sketches require more space to maintain the same level of error bound compared with count sketches. This condition is attributable to the fact that the L_2 norm is generally smaller than the L_1 norm.

In summary, given an input data stream of length N and user specified parameters δ and *epsilon*, the count min sketch technique can store the frequencies of all the instances with the following guarantees: (1) all the stored frequencies differ from the truth at most ϵN with a probability of at least δ; (2) the space usage is $O(\frac{1}{\epsilon} log \frac{1}{\delta})$; and (3) for each update and query, the cost is constant, that is, $O(log \frac{1}{\delta})$.

10.4.5 Some Related Issues on Sketches

In addition to the above introduced sketch based algorithms, several other extended approaches are based on the sketch techniques.

Pseudo random vector generation. A primary issue for sketch construction is that the number of distinct items may be large and, therefore, the size of the corresponding random vector will also be large. This issue will reduce the efficiency of building the sketches. To address this problem, we first generate a set of k random vectors, and then, when the data instance comes in, we can map it to the corresponding pre-generated random vector. This strategy, however, may consume a large amount of space. A more feasible idea is that we can store the random vectors implicitly (i.e., as seeds), which are utilized dynamically to generate the vectors.

The authors in [10] have found that we can generate the random vectors with four-wise independent random vectors from a seed of size $O(log(N))$. Gilbert et al. [37] demonstrated that if Reed-Muller codes are used, we can generate seven-wise independent random vectors. The properties of the pseudo random vectors generation approach are as follows: (1) we can generate a random vector in poly-logarithmic time from the seed; and (2) the dot-product of two vectors can be approximately computed using only their sketch representations. We can observe that the dot product of two vectors is closely related to the Euclidean distance, which is an indication derived through the random projection strategy [45].

Sketch partitioning. Dobra et al. [23] introduced the sketch partitioning technique. The authors deliberately partitioned the join attributes to construct the separate sketches of each group. The final estimation is accumulated from all partitions. The essential part of the introduced technique is the partition of the domains to bind the variance, which can result in high accuracy for applications. The authors in [24] further studied the issue by extending it to multi-query processing.

Sketch skimming. Ganguly et al. [28] observed that sketch skimming can be utilized to improve the estimation of join size. The variance of the join estimation is largely affected by the most frequent random variables, which are generally few even for a large data set. Given that high variance is undesirable, the frequent instances are deliberately separated from others. Therefore, the skimmed sketches can be identified by removing the sketches with frequent instances. We can estimate the join size using four-wise independent random vectors and the experimental evaluation demonstrates the efficiency of the proposed technique in [28].

10.4.6 Applications of Sketches

A large number of applications can utilize the sketch based strategies. One practical issue is the heavy hitters [19, 49, 17]. For this problem, we need to detect the most frequent items in the data stream. Recognizing the difference among networks in the data stream was explored in [20], and detecting the differences among data streams was studied in [25, 26]. Similar issues on XML data (or tree data) were presented in [57, 58, 61]. These works aimed to construct the synopsis for structured queries, which can be used later to improve query processing performance.

Sketches based strategies are also well utilized in network research. Improving the communication efficiency for signals in sensor networks is important. Moreover, considering resource limitations (e.g., battery), efficient storage by using concise summarization on the stream data is an essential issue for sensor networks. A number of works have been conducted to address the aforementioned issues, [15, 38, 46]. Please refer to [8] for more details on these issues.

10.4.7 Advantages and Limitations of Sketch Strategies

Sketch-based strategies have several advantages. First is the space usage. Sketch-based approaches have been theoretically and experimentally proven to obtain an optimal sub-linear space usage in the data size. This finding can be attributed to the fact that the space requirement is logarithmic in the number of distinct items in the stream, which is relevant small by considering the large volume of the data.

Despite the advantages of sketch based methods, several challenging issues remain. First, almost all related studies use L_p norm as the aggregate measure, which may not reflect the actual data distribution. Thus, the sketch summarization may fail to store the essential information of the data.

Another issue is high dimensionality. The existence of hundreds of independent dimensions in the data stream may hinder the practical usage of the existing state-of-the-art sketch-based techniques. This issue has been raised by several researchers [16]. However, the problem remains challenging because of its intrinsic complexity.

As highlighted in [8], most sketch-based works only focus on identifying the frequent instances and estimating the frequency moments and join size. This emphasis on the micro view may neglect the macro trends in the stream data, such as the temporal property. Thus, the temporal information may be lost because of the transformation process when building the sketches. Although several scholars have mentioned this issue and consequently introduced effective techniques for temporal analysis [41], the strategy requires significant space usage, which makes the approach impractical for real large data streams. Extending the state-of-the-art strategies for temporal trends analysis with limited space usage remains an interesting and challenging issue.

As previously mentioned, a trade-off exists between space usage and data rate. For real applications, considering that we have a sufficient storage resource (i.e., main memory, SSD), addressing data streams with relevant slow data update rate (but may have a large volume of distinct items) [37] may be practical. The real challenging applications are therefore those that need to process very fast data streams such as sensor networks [21, 55, 46, 47] because of the power and hardware limitation.

10.5 Histogram Method

The histogram approach is another major tool used for analyzing data streams. In statistics, the term *histogram* was first introduced as a graphical representation to illustrate the visual impression of the distribution of data [2]. A histogram is an estimate of the probability distribution of a continuous variable and was first introduced by Pearson [56].

Histograms are commonly used to describe the density of data and measure the probability density function of the underlying variable. Specifically, in database research, the histogram approach partitions the data into a series of categories (known as bins or buckets) relative to some feature (or dimension). Each count of the bin is stored.

From a formal mathematical view, a histogram is a function m that accumulates the number of instances that fall into each of the disjoint buckets, whereas the graph of a histogram is one method to describe the histogram. Let n be the total number of instances, k be the total number of bins, and m_i be the histogram, we then have $n = \sum_{i=1}^{k} m_i$. A *cumulative histogram* is a mapping that measures the cumulative number of instances in all buckets up to the specified bucket. The cumulative histogram M_i of a histogram m_j is defined as $M_i = \sum_{j=1}^{i} m_j$.

By analyzing the process of histogram construction, we find that the space usage for a histogram is determined by the total number of buckets used. Buckets can intuitively be obtained by partitioning the data into equal sizes. Such equi-width division technique is related to Haar wavelet coefficients in that if the wavelet summarization of the frequency distribution is built relative to any dimension, then the Haar coefficients present the difference in relative frequencies in equi-width histogram buckets [8].

Although this technique is easily implemented for equi-width histogram strategy, it has the drawback of low representation accuracy. This low accuracy can be attributed to the fact that the distribution of data is not well kept by the equi-width mechanism because of the assumption of uniform distribution. The localized data distribution is commonly cut by the bucket boundaries. For instance, the number of points distributed in different buckets may vary significantly. This issue may lend difficulty to query estimations. Therefore, the histogram technique requires the design of an appropriate bucket construction mechanism.

Similar to the idea of *kd*-tree, we can build buckets to enable each one contain approximately equal instances (known as equi-depth histogram). Numerous experiments have illustrated that equi-depth histograms are considerably more effective than equi-width histograms. Therefore, a large number of commercial vendors switched to the equi-depth histograms in the years following their introduction [42]. Multidimensional equi-depth histograms were introduced in [52]. However, for the special data such as a stream, the construction of buckets based on the equi-depth technique is difficult because the data are dynamic and unknown apriori.

To improve the effectiveness of histograms, Ioannidis et al. introduced the *V*-optimal histograms [43], which aim to minimize the frequency variance of different values in buckets. In this way, the assumption of data uniform distribution can be satisfied. Specifically, if a bucket *b* with count *c* contains the frequency of *n* instances, then the average frequency of each instance in *b* is c/n. Let $f_1 \ldots f_n$ be the frequencies of the *n* instances in *b*. The variance *v* of the frequencies based on the averages is obtained as $v = \sum_{i=1}^{l} (f_i - c/n)^2$. Finally, the overall variance *V* on all the buckets is obtained as $V = \sum_b v$.

Improvement on the *V*-optimal histogram construction has been introduced in [44]. In this work, the L_p-difference function between two vectors with cardinalities that are based on the distinct instances is considered as the objective function. Other works consider alternative objective functions to optimize the histogram construction [60]. The advantage and disadvantage of *V*-optimal histograms is explained as follows.

Advantage of *V*-optimal histogram: *V*-optimal histograms can optimally measure the contents of buckets. However, any histogram could encounter an error when used to summarize data. *V*-optimal histograms binds the error by finding the smallest variance among all possible buckets. As demonstrated in [59], *V*-optimal histograms can achieve the best performance in terms of accuracy in summarizing data.

Limitation of *V*-optimal histogram: The major drawback of *V*-optimal histograms is that they are difficult to update. Rebuilding all histograms is necessary when new data are received. By contrast, the equi-width histogram technique can address this issue. Moreover, although equi-depth histograms also have to rebuild, the cost is lower compared with

V-optimal histograms because the structure of the former is simpler and easier to construct. This intrinsic disadvantage may hinder the *V*-optimal histogram from being appropriate strategy for fast dynamical updating of data such as a stream.

Another issue is that numerous studies use absolute errors as the accuracy metric. However, as emphasized by [51], the absolute error may not always be a good representation of the error, thus necessitating the use of other metrics. To address this issue, Guha et al. [40] introduced several strategies to improve the relative error.

Another difference between the equi-width histogram and equi-frequency (*V*-optimal) histogram is that the former would have almost all the samples in one bucket, whereas latter would have numerous narrow buckets in one area even with the same number of buckets. If we consider the height of a bucket as a variable, then the equi-frequency histogram will better spread the available distribution information among the variables.

10.5.1 Dynamic Construction of Histograms

Given the special property of data streams, the requirement of dynamically building the histograms exists for a large number of real applications. In this section, we first review the static histograms and then explore the dynamic histograms. Please refer to [42] for more details.

Static histograms are those that, once built from the original data (or the sample instances), will not change later regardless of whether the original data (or the samples) is changed.

However, as more new data come in (or updated), the error will accumulate until the requirement for query processing applications can no longer be satisfied. To address this issue, recomputing all the histograms is necessary. Therefore, the cost of histogram reconstruction has to be considered as a measure of the performance of different histogram-based algorithms. For a few histogram construction strategies such as equi-width and equi-depth, this factor is not a major problem because the rebuilding process is simple and easy to implement. However, for traditional equi-frequency strategy (i.e., V-optimal histogram), the cost may be high because the number of source parameter values is exponential. Therefore, a trade-off exists between the effectiveness and the efficiency of different histogram construction strategies. To address this issue, dynamic programming-based approaches have been introduced [44, 39]. The work in [44] built V-optimal histograms quadratically based on the number of source parameters and linearly based on the number of buckets. This contribution makes the V-optimal histograms acceptable for the histogram rebuilding scenario. The work in [39] reduced the total cost to be linear to the number of source parameters. Despite these achievements, however, building Voptimal histograms on multi-dimension data. Thus, [54] introduced the approximate strategies.

For dynamic data such as a stream, the aforementioned techniques may be ineffective because the data in the stream can be scanned only once, and the introduced strategies always need to verify the data multiple times. Several studies have been conducted to address this issue. For example, Gibbons et al. [33] proposed the equi-depth histograms based approach. Gilbert et al. [34] introduced the V-optimal histogram based technique for data stream processing. Given the high complexity and importance of this issue, histogram construction for data streams remains an open and challenging topic in the literature.

10.6 Discussion

A number of challenging issues should be addressed in future research on building synopses for data streams.

- Comparing different kinds of synopsis-based strategies such as sampling, wavelet, sketch, and histogram remains difficult. Different techniques may have their own advantage for specific applications yet may lose their effectiveness when employed for other applications. Thus, comprehensive comparisons among these approaches are necessary. For fair assessment, different setting environments have to be built for the evaluation of the performances of the strategies relative to effectiveness in terms of the error bound, efficiency in terms of synopsis construction and consumed space, as well as usage on high-dimensional data streams. Furthermore, analyses must be conducted not only from a micro view (e.g., frequent item counting), but also from a macro view (e.g., temporal trend detection).
- Workload aware strategy is one of the possible ways to improve the efficiency and effectiveness of synopsis construction. Several groups have already studied this issue [54, 50]. However, the complex dynamic properties of data streams require more intelligent techniques to provide higher effectiveness with lower cost of synopsis construction and update.
- Considering that the current data type taken into account is commonly quantitative or categorical, the future direction is to extend the data type to others, e.g., text, XML, and so on. Some studies, such as [58, 57], have already addressed this issue yet more researches are preferred. We believe that there is considerable scope for extension of the current synopsis methods to domains such as sensor data mining in which the hardware requirements force the use of space-optimal synopsis. However, the objective of constructing a given synopsis needs to be

carefully calibrated in order to take the specific hardware requirements into account. While the broad theoretical foundations of this field are now in place, it remains to carefully examine how these methods may be leveraged for applications with different kinds of hardware, computational power, or space constraints.

10.7 Chapter Summary

In this chapter, we presented an overview of the different methods to construct synopsis for data streams. We introduced random sampling, wavelets, sketches, and histograms. In addition to the properties of different strategies, the advantages and limitations of these approaches have been thoroughly discussed. We also gave some possible challenges which may be the future works explored in the literature.

References

[1] http://en.wikipedia.org/wiki/Simple_random_sample.
[2] http://en.wikipedia.org/wiki/Histogram.
[3] Cluster sampling explanation on wiki, 2012.
[4] Sampling explanation on wiki, 2012.
[5] Stratified sampling explanation on wiki, 2012.
[6] Wavelet explanation on wiki, 2012.
[7] K. D. A. and H. M. Wavelets and their applications in databases. *In: Proceedings of the 21st International Conference on Data Engineering*, 2001.
[8] C. Aggarwal, editor. *Data Streams—Models and Algorithms*. Springer, 2007.
[9] C. C. Aggarwal, J. Han, J. Wang and P. S. Yu. A framework for clustering evolving data streams. *In: Proceedings of the 29th International Conference on Very Large Data bases*, Vol. 29, pp. 81–92, 2003.
[10] N. Alon, Y. Matias and M. Szegedy. The space complexity of approximating the frequency moments. *In: Proceedings of the Twenty-eighth Annual ACM Symposium on Theory of computing*, pp. 20–29, 1996.
[11] B. Babcock, M. Datar and R. Motwani. Sampling from a moving window over streaming data. *In: Proceedings of the Thirteenth Annual ACM-SIAM Symposium on Discrete algorithms*, pp. 633–634, 2002.
[12] C. M. Bishop. *Pattern Recognition and Machine Learning (Information Science and Statistics)*. Springer-Verlag New York, 2006.
[13] V. Braverman, R. Ostrovsky and C. Zaniolo. Optimal sampling from sliding windows. *In: Proceedings of the twenty-eighth ACM SIGMOD-SIGACT-SIGART Symposium on Principles of database systems*, pp. 147–156, 2009.
[14] M. Charikar, K. Chen and M. Farach-Colton. Finding frequent items in data streams. *In: Proceedings of the 29th International Colloquium on Automata, Languages and Programming*, pp. 693–703, 2002.
[15] G. Cormode and M. Garofalakis. Sketching streams through the net: distributed approximate query tracking. *In: Proceedings of the 31st International Conference on Very Large Data Bases*, pp. 13–24, 2005.
[16] G. Cormode, M. N. Garofalakis and D. Sacharidis. Fast approximate wavelet tracking on streams. *In: EDBT*, pp. 4–22, 2006.

[17] G. Cormode and M. Hadjieleftheriou. Finding frequent items in data streams. *Proc. VLDB Endow.*, 1: 1530–1541, August 2008.

[18] G. Cormode and S. Muthukrishnan. An improved data stream summary: the count-min sketch and its applications. *Journal of Algorithms*, 55(1): 58–75, Apr. 2005.

[19] G. Cormode and S. Muthukrishnan. What's hot and what's not: tracking most frequent items dynamically. *ACM Trans. Database Syst.*, 30: 249–278, March 2005.

[20] G. Cormode and S. Muthukrishnan. What's new: finding significant differences in network data streams. *IEEE/ACM Trans. Netw.*, 13: 1219–1232, December 2005.

[21] A. Das, S. Ganguly, M. Garofalakis and R. Rastogi. Distributed set-expression cardinality estimation. *In: Proceedings of the Thirtieth International Conference on Very Large Data Bases* —Vol. 30, pp. 312–323, 2004.

[22] A. Deligiannakis, M. Garofalakis and N. Roussopoulos. Extended wavelets for multiple measures. *ACM Trans. Database Syst.*, 32, June 2007.

[23] A. Dobra, M. Garofalakis, J. Gehrke and R. Rastogi. Processing complex aggregate queries over data streams. *In: Proceedings of the 2002 ACM SIGMOD International Conference on Management of Data*, SIGMOD'02, pp. 61–72, 2002.

[24] A. Dobra, M. Garofalakis, J. Gehrke and R. Rastogi. Sketch-based multi-query processing over data streams. *In: EDBT*, pp. 551–568, 2004.

[25] J. Feigenbaum, S. Kannan, M. Strauss and M. Viswanathan. An approximate l1-difference algorithm for massive data streams. *In: Proceedings of the 40th Annual Symposium on Foundations of Computer Science*, 1999.

[26] J. H. Fong and M. Strauss. An approximate lp-difference algorithm for massive data streams. *In: Proceedings of the 17th Annual Symposium on Theoretical Aspects of Computer Science*, pp. 193–204, 2000.

[27] M. M. Gaber, A. Zaslavsky and S. Krishnaswamy. Mining data streams: a review. *SIGMOD Rec.*, 34(2): 18–26, Jun. 2005.

[28] S. Ganguly, M. Garofalakis and R. Rastogi. Processing data-stream join aggregates using skimmed sketches. *In: Proc. Int. Conf. on Extending Database Technology*, pp. 569–586, 2004.

[29] M. Garofalakis and P. B. Gibbons. Wavelet synopses with error guarantees. *In: Proceedings of the 2002 ACM SIGMOD international conference on Management of data*, pp. 476–487, 2002.

[30] M. Garofalakis and A. Kumar. Deterministic wavelet thresholding for maximum-error metrics. *In: Proceedings of the twenty-third ACM SIGMOD-SIGACT-SIGART Symposium on Principles of Database Systems*, pp. 166–176, 2004.

[31] M. N. Garofalakis. Wavelets on streams. *In: Encyclopedia of Database Systems*, pp. 3446–3451. 2009.

[32] R. Gemulla and W. Lehner. Sampling time-based sliding windows in bounded space. *In: Proceedings of the 2008 ACM SIGMOD International Conference on Management of data*, pp. 379–392, 2008.

[33] P. B. Gibbons, Y. Matias and V. Poosala. Fast incremental maintenance of approximate histograms. *ACM Trans. Database Syst.*, 27(3): 261–298, Sep. 2002.

[34] A. C. Gilbert, S. Guha, P. Indyk, Y. Kotidis, S. Muthukrishnan and M. J. Strauss. Fast, small-space algorithms for approximate histogram maintenance. *In: Proceedings of the Thiry-fourth Annual ACM Symposium on Theory of Computing*, pp. 389–398, 2002.

[35] A. C. Gilbert, Y. Kotidis, S. Muthukrishnan and M. Strauss. Surfing wavelets on streams: One-pass summaries for approximate aggregate queries. *In: Proceedings of the 27th International Conference on Very Large Data Bases*, pp. 79–88, 2001.

[36] A. C. Gilbert, Y. Kotidis, S. Muthukrishnan and M. Strauss. Surfing wavelets on streams: One-pass summaries for approximate aggregate queries. *In: Proceedings of the 27th International Conference on Very Large Data Bases*, pp. 79–88, 2001.

[37] A. C. Gilbert, Y. Kotidis, S. Muthukrishnan and M. Strauss. Surfing wavelets on streams: One-pass summaries for approximate aggregate queries. *In: Proceedings of the 27th International Conference on Very Large Data Bases*, pp. 79–88, 2001.

[38] M. B. Greenwald and S. Khanna. Power-conserving computation of order-statistics over sensor networks. *In: Proceedings of the twenty-third ACM SIGMOD-SIGACT-SIGART Symposium on Principles of Database Systems*, pp. 275–285, 2004.

[39] S. Guha, P. Indyk, S. Muthukrishnan and M. Strauss. Histogramming data streams with fast per-item processing. *In: Proceedings of the 29th International Colloquium on Automata, Languages and Programming*, pp. 681–692, 2002.

[40] S. Guha, K. Shim and J. Woo. Rehist: relative error histogram construction algorithms. *In: Proceedings of the Thirtieth International Conference on Very large data bases*—Vol. 30, pp. 300–311, 2004.

[41] P. Indyk, N. Koudas and S. Muthukrishnan. Identifying representative trends in massive time series data sets using sketches. *In: Proceedings of the 26th International Conference on Very Large Data Bases*, VLDB '00, pp. 363–372, 2000.

[42] Y. Ioannidis. The History of Histograms (abridged). *In: Proceedings of the 29th International conference on Very Large Data Bases*—Vol. 29, pp. 19–30, 2003.

[43] Y. E. Ioannidis and V. Poosala. Balancing histogram optimality and practicality for query result size estimation. *In: Proceedings of the 1995 ACM SIGMOD International Conference on Management of Data*, pp. 233–244, 1995.

[44] H. V. Jagadish, N. Koudas, S. Muthukrishnan, V. Poosala, K. C. Sevcik and T. Suel. Optimal Histograms with Quality Guarantees. *In: Proceedings of the 24th International Conference on Very Large Data Bases*, pp. 275–286, 1998.

[45] W. B. Johnson and J. Lindenstrauss. Extensions of Lipschitz mapping into Hilbert space. *In: Conf. in Modern Analysis and Probability*, Vol. 26, pp. 189–206, 1984.

[46] D. Kempe, A. Dobra and J. Gehrke. Gossip-based computation of aggregate information. *In: Proceedings of the 44th Annual IEEE Symposium on Foundations of Computer Science*, 2003.

[47] G. Kollios, J. W. Byers, J. Considine, M. Hadjieleftheriou and F. Li. Robust aggregation in sensor networks. *IEEE Data Eng. Bull.*, 28(1): 26–32, 2005.

[48] S. L. Lohr. *Sampling: Design and Analysis*. Duxbury Press, Dec 1999.

[49] G. S. Manku and R. Motwani. Approximate frequency counts over data streams. *In: Proceedings of the 28th International Conference on Very Large Data Bases*, pp. 346–357, 2002.

[50] Y. Matias and D. Urieli. Optimal workload-based weighted wavelet synopses. *In: ICDT*, pp. 368–382, 2005.

[51] Y. Matias, J. S. Vitter and M.Wang. Wavelet-based histograms for selectivity estimation. *In: Proceedings of the 1998 ACM SIGMOD International Conference on Management of data*, pp. 448–459, 1998.

[52] M. Muralikrishna and D. J. DeWitt. Equi-depth histograms for estimating selectivity factors for multi-dimensional queries. *In: Proceedings of the 1988 ACM SIGMOD International Conference on Management of Data*, pp. 28–36, 1988.

[53] S. Muthukrishnan. Data streams: Algorithms and applications. *Foundations and Trends in Theoretical Computer Science*, 1(2), 2005.

[54] S. Muthukrishnan, V. Poosala and T. Suel. On rectangular partitionings in two dimensions: Algorithms, complexity, and applications. *In: Proceedings of the 7th International Conference on Database Theory*, pp. 236–256, 1999.

[55] C. Olston, J. Jiang and J.Widom. Adaptive filters for continuous queries over distributed data streams. *In: Proceedings of the 2003 ACM SIGMOD international conference on Management of data*, pp. 563–574, 2003.

[56] K. Pearson. Contributions to the mathematical theory of evolution. ii. skew variation in homogeneous material. *Philosophical Transactions of the Royal Society A: Mathematical, Physical and Engineering Sciences*, (186): 326–343, 1895.

[57] N. Polyzotis and M. Garofalakis. Structure and value synopses for xml data graphs. *In: Proceedings of the 28th International Conference on Very Large Data Bases*, pp. 466–477, 2002.

[58] N. Polyzotis and M. Garofalakis. Xcluster synopses for structured xml content. *In: Proceedings of the 22nd International Conference on Data Engineering*, 2006.

[59] V. Poosala, P. J. Haas, Y. E. Ioannidis and E. J. Shekita. Improved histograms for selectivity estimation of range predicates. *In: Proceedings of the 1996 ACM SIGMOD International Conference on Management of Data*, pp. 294–305, 1996.

[60] V. Poosala and Y. E. Ioannidis. Selectivity estimation without the attribute value independence assumption. *In: Proceedings of the 23rd International Conference on Very Large Data Bases*, pp. 486–495, 1997.

[61] P. Rao and B. Moon. Sketchtree: Approximate tree pattern counts over streaming labeled trees. *In: Proceedings of the 22nd International Conference on Data Engineering*, 2006.

[62] F. Rusu and A. Dobra. Statistical analysis of sketch estimators. *In: Proceedings of the 2007 ACM SIGMOD International Conference on Management of Data*, pp. 187–198, 2007.

[63] E. J. Stollnitz, T. D. Derose and D. H. Salesin. *Wavelets for Computer Graphics: Theory and Applications*. Morgan Kaufmann Publishers Inc., 1996.

[64] M. Thorup and Y. Zhang. Tabulation based 4-universal hashing with applications to second moment estimation. *In: Proceedings of the fifteenth annual ACM-SIAM Symposium on Discrete Algorithms*, pp. 615–624, 2004.

[65] J. S. Vitter. Random sampling with a reservoir. *ACM Trans. Math. Softw.*, 11(1): 37–57, Mar. 1985.

[66] J. S. Vitter and M. Wang. Approximate computation of multidimensional aggregates of sparse data using wavelets. *In: Proceedings of the 1999 ACM SIGMOD International Conference on Management of Data*, pp. 193–204, 1999.

[67] D. S. Yates, D. S. Moore and D. S. Starnes. *The Practice of Statistics*. Freeman, 2008.

[68] H. A. Zur. Theorie der orthogonalen funktionensystemes. *Mathematische Annalen*, (69): 331–371, 1910.

CHAPTER 11

Recommendation Systems

Recommendation systems are important applications that are essential for numerous business models. Recommendation systems suggest appropriate items based on user preference and historical purchase data. These systems are based on the principle that if users shared the same interests in the past, they will, with high probability, exhibit similar behavior in the future. The historical data that reflect user preference may comprise explicit ratings, Web click logs, or tags. Personalization is evidently an important factor in an effective recommendation system. In this chapter, we will introduce the basic concepts and main strategies for recommendation systems.

The collaborative filtering (CF) approach will be presented in Section 11.1, in which user- and item-based CF methods are introduced. The probability latent semantic analysis (PLSA) will be presented in Section 11.2. The tensor method will be introduced in Section 11.3. A discussion on data stream will be presented in Section 11.4. A chapter summary will be given in Section 11.5.

11.1 Collaborative Filtering

One of the most basic and important techniques in recommendation systems is collaborative filtering (CF). The key idea of CF is that automatic predictions (or filtering) are made about the interests of users by collecting preference information from a large number of users (i.e., collaborate). The preference data may include explicit ratings, Web click logs, reviews, or tags. Through deliberate analysis of the interrelation between people (represented as user profile) and items based on preference information, effective recommendations can be suggested. To encode the profile of a user, a common method is to use a vector of the user's ratings on items. The rating values can be either binary (i.e., like or dislike) or numeric values that indicate the degree of the rating. Researchers have proposed two categories of CF algorithms: memory- and model-based [17, 3, 11]. We will introduce these two kinds of methods in the subsequent sections.

11.1.1 Memory-based Collaborative Recommendation

Memory-based collaborative methods always employ the total ratings of users in the training data to make a recommendation. These strategies can be further divided into two classes: user- and item-based approaches [20].

11.1.1.1 User-based Recommendation

In this section, we introduce one representative user-based recommendation algorithm: the user-based k nearest neighborhood algorithm (UBkNN). UBkNN finds a set of users who have similar preferences as the target user by calculating the similarity among users. To fulfill this purpose, the algorithm applies a number of state-of-the-art kNN (i.e., top-k nearest neighborhood) strategies. After the process of finding the kNN users, the approach applies the common CF algorithm to propose a list of item recommendations to the user. Given a query user u, the recommendation of item i is computed as follows: $p_{u,i} = \frac{\sum_{j=1}^{k}(R_{j,i}\,sim(u,j))}{\sum_{j=1}^{k}sim(u,j)}$, where $R_{j,i}$ denotes the rating by user j on item i, whereas the k most similar users (with regard to user i) are considered.

11.1.1.2 Item-based Recommendation

In contrast to the UBkNN algorithm, the item-based kNN method [20] is another kind of CF approach that computes the similarity between two items, instead of users. In the item-based kNN algorithm, the similarity among items is computed by comparing the item vector, after which a similarity table is constructed. In this table, each row is modeled as a set of ratings by all users on one item, whereas each column is modeled as a set of ratings by one user on all the items. To assess the rating on an item i for user u, the algorithm computes the ratio of the sum of the ratings given by the user on the items that are similar to i with respect to the sum of involved item similarities as follows: $p_{u,i} = \frac{\sum_{j=1}^{k}(R_{u,j}\,sim(i,j))}{\sum_{j=1}^{k}sim(i,j)}$, where $R_{u,j}$ denotes rating by user u on item j, and the k most similar items (with regard to item i) are considered.

11.1.2 Model-based Recommendation

Another main class of CF algorithms is the model-based approach. Model-based recommendation constructs a model from the historical data (i.e., rating, tag, etc.) and then uses this model to make a recommendation. Several approaches can be used to build the model, such as the hidden Markov model, decision tree, clustering, Bayesian networks, neural networks, latent semantic analysis, and so on.

Mobasher et al. introduced a model-based recommendation system, i.e., Profile Aggregations based on Clustering Transaction that applies clustering strategies to aggregate user sessions. Users are then clustered based on similar preferences (i.e., access pattern). The clustering model learned from the training data can be used to make a recommendation for a newcomer, such that the representative of the clustering is suggested to the person who has a similar access pattern as the cluster.

Notably, the similarity metric is important for its function in evaluating how similar two users (or items) are. Common metrics include cosine, jaccard, and so on. However, this subject is not within the scope of this chapter, and we direct interested users to [20].

11.2 PLSA Method

The PLSA model was first proposed in [10] to address text mining. The basic idea of PLSA is related to that of LSA [5], the difference being the fact that the latter is based on linear algebra and downsizes the occurrence tables (via a singular value decomposition), whereas the former is built by mixture decomposition derived from a latent class model in statistic theory [1]. PLSA intuitively aims to recognize the hidden semantic relationships among co-occurrence activities, usually based on the aspect model.

To illustrate the PLSA, we present its application on Web usage mining. User sessions over Web pages can be deemed as co-occurrence activities to deduce the latent usage pattern. The aspect model assumes the existence of a latent factor space $Z = (z_1, z_2, \ldots, z_k)$, and each co-occurrence observation data (s_i, p_j) (i.e., the visit of page p_j in user session s_i) is associated with the factor $z_k \in Z$ by a varying degree to z_k. Intuitively, the relationships between users and Web pages should be different and determined by a variety of factors, which can then be used to represent the latent usage patterns of the users.

For example, when applying PLSA on an e-shopping website, we can assume k categories of navigational behavior patterns (determined by k latent factors). The k factors could be the probabilities that: (1) users have an interest in the travel-related product category; (2) users merely browse different products; (3) users tend to buy entertainment products, and so on.

To reflect all of these probabilities, we can project the training data into the corresponding latent factor space. The representation of these projections can be defined as the conditional probability distribution that reflects the relationships among users or Web pages (which are, indeed, latent usage patterns). In a brief summary, PLSA aims to identify and represent user access behavior in latent semantic spaces, and determine the corresponding factors. In the following section, the mathematical theory of PLSA will be presented. First, we give several definitions that are necessary in the framework: $P(s_i)$ represents the probability that a user session s_i appears in the training data; $P(z_k | s_i)$ indicates that, given a user session s_i, the probability of the latent factor z_k associated with s_i; and $P(p_j | z_k)$ denotes that, given the latent factor z_k, the probability of the pages p_j exists.

The algorithm of the PLSA model is executed as the following steps: (1) a user session s_i is selected with probability $P(s_i)$; (2) a factor z_k is chosen with probability $P(z_k | s_i)$; and (3) a Web page p_j is presented with probability $P(p_j | z_k)$. Through these steps, we can derive the probability of the observation data (s_i, p_j) relative to the latent factor z_k. The process can be presented by the following formula:

$$P(s_i, p_j) = P(s_i) \cdot P(p_j | s_i) \tag{11.2.1}$$

where, $P(s_i, p_j) = P(s_i) \cdot P(p_j | s_i)$. Through the Bayesian rule, the above equations can be transformed to:

$$P(s_i, p_j) = \sum_{z \in Z} P(z)P(s_i | z)P(p_j | z) \tag{11.2.2}$$

Based on the likelihood rule, the total likelihood of the observation data can be presented as:

$$L_i = \sum_{s_i \in S, p_j \in P} m(s_i, p_j) \cdot log P(s_i, p_j) \tag{11.2.3}$$

where $m(s_i, p_j)$ denotes the element of the matrix (determined by user sessions and web pages) associated with user session s_i and page access p_j.

To maximize the total likelihood, the conditional probabilities $P(z)$, $P(s_i | z)$, and $P(p_j | z)$ must be recursively optimized based on the observation data. The Expectation Maximization (*EM*) strategy [6] is known to be an effective tool for addressing this issue. In *EM*, two steps are recursively implemented: (1) expectation (*E*) step, where the posterior probabilities are computed for the latent factors based on the current computations of the conditional probability; and (2) maximization (*M*) step, where the estimated conditional probabilities are updated and used to maximize the likelihood based on the posterior probabilities computed in the previous *E* step.

The procedure is executed as follows: We first set the initial values of $P(z)$, $P(s_i | z)$, and $P(p_j | z)$ randomly. In the *E*-step, we employ the Bayesian rule to compute the following values relative to the observation data:

$$P(z_k \mid s_i, p_j) = \frac{P(z_k)P(s_i \mid z_k)P(p_j \mid z_k)}{\sum_{z_k \in Z} P(z_k)P(s_i \mid z_k)P(p_j \mid z_k)} \tag{11.2.4}$$

In the M-step, we calculate the following values:

$$P(p_j \mid z_k) = \frac{\sum_{s_i \in S} m(s_i, p_j)P(z_k \mid s_i, p_j)}{\sum_{s_i \in S, p'_j \in P} m(s_i, p'_j)P(z_k \mid s_i, p'_j)} \tag{11.2.5}$$

$$P(s_i \mid z_k) = \frac{\sum_{p_j \in P} m(s_i, p_j)P(z_k \mid s_i, p_j)}{\sum_{s'_i \in S, p_j \in P} m(s'_i, p_j)P(z_k \mid s'_i, p_j)} \tag{11.2.6}$$

$$P(z_k) = \frac{1}{R} \sum_{s_i \in S, p_j \in P} m(s_i, p_j)P(z_k \mid s_i, p_j) \tag{11.2.7}$$

where $R = \sum_{s_i \in S, p_j \in P} m(s_i, p_j)$. Substituting Eqs. 11.2.5 with 11.2.7 into Eqs. 11.2.2 to 11.2.3 will yield the total likelihood L_i of the observation data with monotonic increasing property. The E-step and M-step are recursively executed until convergence occurs, which indicates that the result is maximized to be the optimal estimate of the observation data. In terms of the complexity of the PLSA algorithm, the computational cost is $O(mnk)$, where m denotes the number of sessions, n denotes the number of the Web pages, and k denotes number of latent factors.

Through the aforementioned process, we can see that the estimated probability distribution intrinsically reflects the local maximum likelihood and therefore encodes the critical information that could be used to deduce the latent factors.

11.2.1 User Pattern Extraction and Latent Factor Recognition

In the PLSA model, latent factors (are assumed to) indicate features that reflect usage co-occurrence observation activities. As an intuitive result, every latent factor could have a specific user access pattern. To address the issue of decoding latent factor and extracting user patterns, we can build aggregated user profiles to present the user access behaviors based on the estimated probability distributions. A simple representation for the aggregated user profiles is achieved by using a set of clustered pages that are weighted to illustrate their contributions to the clustered group. The semantic meaning of the latent factor can be deduced by analyzing the aggregated user profile, that is, the representative topic of the cluster group.

11.2.1.1 User Session Partition

Given a user session s_i, the estimated probability distribution in the factor space may indicate the user's access pattern over the whole latent factor space, which can thus be explored to discover the dominant factors by recognizing the top probability values. By using Bayesian rule, we can calculate a set of probabilities over the latent factor space as follows:

$$P(z_k \mid s_i) = \frac{P(s_i \mid z_k)P(z_k)}{\sum_{z_k \in Z} P(s_i \mid z_k)P(z_k)} \tag{11.2.8}$$

Considering that only a few probability distributions can pass the predefined threshold test, the probability group $P(z_k \mid s_i)$ is always very sparse. To mitigate this problem, the users can be clustered into a corresponding probability distribution that is larger than the threshold. Notably, a user session can be represented by a set of pages, and a mixture model can be utilized based on the latent factor z_k relative to the weighted pages. The pseudo code of user session partition is shown as follows:

Algorithm 6: User Session Partition

Input: A set of calculated probability values of $P(z_k|s_i)$, a user session-page matrix SP, and a predefined threshold μ.
Output: A set of session clusters $SCL=(SCL_1, SCL_2, \ldots SCL_k)$
Set $SCL_1 = SCL_2 = \ldots = SCL_k = \varphi$;
for *each* $s_i \in S$ **do**
    select $P(z_k|s_i)$;
    if $P(z_k|s_i) \geq \mu$ **then**
        $SCL_k = SCL_k \cup s_i$;
    end
end
if *there are remaining users sessions to be clustered* **then**
    go to line 2;
end
Output session clusters $SCL = \{ SCL_k \}$;

11.2.1.2 Latent Factor Recognition

Analyzing the latent factor is important because of its significance in the PLSA model. To address this issue, similar to the user session partition, the probability distribution can be employed to partition Web pages into corresponding clusters relative to the latent factors. A threshold-based strategy can also be used to identify the conditional probabilities that pass the test and possess similar semantic meaning. After clustering, the URLs of the pages and the weights deduced from the model will be utilized to analyze the semantic meaning of the latent factors. The pseudo code of the algorithm for recognizing latent factors is as follows:

Algorithm 7: Latent Factor Recognition

Input: A set of conditional probabilities, $P(p_j|z_k)$, a predefined threshold μ
Output: A set of latent semantic factors represented by several essential pages
Set $PCL_1 = PCL_2 = \ldots = PCL_k = \varphi$;
for *each* z_k **do**
$\quad$ select the web pages which have $P(p_j|z_k) \geq \mu$ and $P(z_k|p_j) \geq \mu$;
$\quad$ $PCL_k = p_j \cup PCL_k$;
end
if *there are remaining users pages to be clustered* **then**
$\quad$ go to line 2;
end
Output $PCL = \{ PCL_k \}$;

11.3 Tensor Model

Tensors are geometric objects that describe linear relations among vectors, scalars, and other tensors [2]. In this section, we will briefly introduce the tensor method, which is a commonly used strategy for recommendation systems.

A matrix is an effective tool that encodes the relationship between two types of objects, such as the information between the users and their clicked Web pages. A common characteristic of a matrix is that each row can be considered as a linear combination of values from different column spaces, and vice versa, where each column is represented by a vector of elements in the row space. Computation based on matrix can effectively address a number of real problems because two dimensional model (i.e., matrix-based model) can fit these problems well. Nevertheless, high-dimensional problems such as user vs. pages vs. time vs. keywords must likewise be addressed. A tensor, which can be considered as a high-dimensional version of a matrix, can be considered as a general model for high-dimensional data. Therefore, the tensor model is employed for all problems that involve multiple dimensional issues. The existence of numerous models related to tensor also provides a powerful tool. We discuss the mathematical background of the tensor model as follows:

We first present a number of basic definitions used in the tensor model with meanings that differ from those under a two-dimensional situation. Specifically, the order, mode, and dimension are used to denote the concepts of dimensionality, dimension, and attribute value that are used in linear algebra. For instance, a third-order tensor is the same as a three-dimensional data expression. Furthermore, we define several specific symbols for the tensor model, which are presented as follows: (1) *scalar* is denoted by a lowercase letter, such as a; (2) *vector* is denoted by a boldface lowercase letter, such as **a**; (3) the *i*th entry of **a** is denoted by $\mathbf{a}_i$; (4) *matrix* is denoted by a boldface capital letter, such as **A**; (5) the *j*-th column of **A** is denoted by $\mathbf{a}_j$, whereas the element of *j*th column and *i*th row is denoted by $\mathbf{a}_{ij}$; (6) *tensor* is denoted by an italicized boldface letter, such as *X*; (7) element (*i*, *j*, *k*) of a third-order tensor is denoted by X_{ijk}; and (8) a tensor of order *M* closely resembles a data cube with *M* dimensions. Formally, we write an *M*th order tensor $X \in R^{N_1 \times N_2 \times \dots N}{}_m$, where $N_i (1 \leq i \leq M)$ is the dimensionality of the *i*th mode. For brevity, we often omit the subscript $[N_1, \dots, N_M]$. Moreover, more important concepts used in the tensor model are defined as follows [21].

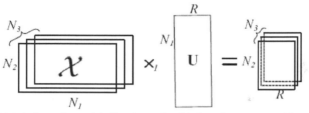

Figure 11.3.1: Sample multiplication of a third-order tensor with a matrix

Definition 2 (Matricizing or Matrix Unfolding) [21]. The mode-*d* matricizing or matrix un-folding of an *M*th order tensor $X \in R^{N_1 \times N_2 \times \dots N}{}_m$ *is a vector in* $R_N d$ *obtained by keeping index d fixed and varying the other indices. Therefore, the mode-d matricizing X(d) is in* $R^{\Pi_{i \neq d} N_i \times N_d}$.

Definition 3 (Mode Product) [21]. The mode product $X \times_d U$ of a tensor $X \in R^{N_1 \times N_2 \times \dots N}{}_m$ and a matrix $U \in R^{N_d \times N'}$ is the tensor in $R^{N_1 \times \dots \times N_{d-1} \times N' \times N_{d+1} \times \dots \times N_M}$ defined by:

$$X \times_d U(i_1, \dots, i_{d-1}, j, i_{d+1}, \dots, i_M) = \sum_{i_d=1}^{N_i} X(i_1, \dots, i_{d-1}, i_d, i_{d+1}, \dots, i_M) U(i_d, j) \qquad (11.3.1)$$

for all index values.

An example is illustrated in Fig. 11.3.1 for third-order tensor **X** (i.e., three-dimensional data) mode-1 multiplied by a matrix **U**. The process is executed in three steps: (1) matricizing **X** along mode-1; (2) performing matrix multiplication between $\times_1$ and **U**; and (3) folding the result back as a tensor.

Based on Definition 2, we can calculate a set of multiplications of a tensor $X \in R^{N_1 \times N_2 \times \dots N}{}_m U_i |{}^M_{i=1} \in R^{N_i \times D_i}$ as: $\mathbf{X} \times_1 U_1 \dots \times_m U_M \in R^{D_1 \times \dots \times D_M}$, which can be represented as $\times \sum_{i=1}^M \times_i U_i$. Moreover, we present the following multiplications of all U_j except the *i*-th: $X \times_1 U_1 \dots \times_{i-1} U_{i-1} \times_{i+1} U_{i+1} \dots \times_M U_M$ as $X \Pi_{j \neq i} \times_j U_j$.

Definition 4 (Rank-$(R_1, \ldots, R_M)$ approximation). Given a tensor $X \in R^{N_1 \times \ldots N_M}$, its best Rank-$D_1, \ldots, D_M$ approximation is the tensor $\tilde{X} \in R^{D_1 \times \ldots \cdot D_M}$ with rank $\tilde{X}(d) = D_d$ for $1 \leq d \leq M$, which satisfies the optimal criterion of least square error argmin $||X - \tilde{X}||^2_F$.

The best Rank-$(R_1, \ldots, R_M)$ approximation is $\tilde{X} = Y \Pi_{j=1}^M \times_j U_j$, where the tensor Y is the core tensor of approximation $Y \in R^{N_1 \times \ldots \times N_M}$, and $U_j |_{j=1}^M \in R^{N_j \times D_j}$ is the projection matrices.

11.4 Discussion and Challenges

Recommendation systems are confronted by several issues. The first issue is the cold start problem, which refers to a case in which items (or users) that are not rated by others (or new user) are not recommended. Numerous studies have been conducted to address this issue. Another challenge is the sparsity issue, a case in which only a small percentage of the total items are rated by users [15]. To address the sparsity issue, several works have introduced an award-giving mechanism that encourages users to rate more items. Other works focus on the implicit behavior of users, which indicates the users' rating [18]. Other problems for recommendation systems include data redundancy, noisy data, and so on [23].

In addition to the aforementioned problems, we will introduce other important issues related to the recommendation systems.

11.4.1 Security and Privacy Issues

A recommendation system is known to achieve optimal performance when more information is known about the users, which means that the users need to present as much personal information as possible to the system to obtain good suggestions. This process, however, may give rise to privacy problem. Personal information typically includes the user's name, birth date, postal code, email, and so on. A registration process is always necessary if a user hopes to obtain a recommendation from the system. As explained in Chapter 9 (i.e., issues on privacy preservation), combinations of such personal information may be highly identifying (Quasi-identifier[1]). Therefore, the personal data submitted to the recommendation systems may become quasi-identifiers [12]. Moreover, such personal information may be disseminated, intended or unintended, by the recommendation system.

---

[1] Quasi-identifier: "Variable values or combinations of variable values within a dataset that are not structural uniques but might be empirically unique and therefore in principle uniquely identify a population unit."(OECD, Glossary of statistical term, 2010)

In an ideal environment, users should trust that not only will recommendation systems protect their privacy, but will also provide highly accurate resultant recommendations [12]. Nevertheless, this condition is not true for numerous real applications.

Considering these problems, recommendation systems should prevent the disclosure or misuse of user' data. Other security-related issues also exist. For instance, a product creator may manipulate the recommendation provided by the system such that his product will be recommended to users [4, 13].

11.4.2 Effectiveness Issue

Recommendation systems primarily aim to provide good suggestions to users. This aim embodies the effectiveness issue. The evaluation of effectiveness has thus been an important and thoroughly studied subject over the past several decades [8, 9, 14, 22]. A large number of commonly used evaluation metrics are based on *coverage* and *accuracy*. Coverage estimates the percentage of items that a recommendation system can recommend [8]. Accuracy can be calculated through statistical or decision support-based methods [8].

Statistics-based metrics include root mean squared error, mean absolute error, and so on. The basic idea for statistics-based metrics is that computed ratings are compared with real ratings. Decision support-based metrics include those commonly used in the information retrieval literature, such as precision (the percentage of real "high" ratings compared with those computed to be "high" by recommendation systems), recall (the percentage of real computed to be "high" ratings compared with those known to be "high"), F measure, and so on [8]. Support-based metrics compute how well recommendation systems make suggestions.

Despite the given metrics, tests of recommendation effectiveness on an unbiased random sample remain limited because uncovering the real scenario is time consuming [14]. Thus, existing experimental evaluations only test data that users have already selected to rate, which may introduce bias, that is, users may rate mostly the items that they like. Moreover, for real recommendation systems, relying solely on accuracy, recall, or any of the given metrics is impractical. For instance, in a supermarket application, recommending obvious items (e.g., via the association rule) will yield high precision but may not be helpful to the user because the user is already familiar with such items. Thus, recommendation systems must provide uncommon and useful recommendations based on economics-oriented measures, similar to those given [7, 16, 19].

11.5 Chapter Summary

In this chapter, we provided an overview of the basic concepts and different methods for recommendation systems. We discussed the CF, PLSA, and tensor methods. In addition, we discussed the important issues and problems related to recommendation systems, including cold start, data sparsity, privacy, and effectiveness. This chapter explored the basic methodologies that could be further explored for interested readers.

References

[1] http://en.wikipedia.org/wiki/Probabilistic_latent_semantic_analysis.

[2] http://en.wikipedia.org/wiki/Tensor.

[3] J. S. Breese, D. Heckerman and C. M. Kadie. Empirical analysis of predictive algorithms for collaborative filtering. *In: UAI*, pp. 43–52, 1998.

[4] P.-A. Chirita, W. Nejdl and C. Zamfir. Preventing shilling attacks in online recommender systems. *In: Proceedings of the 7th annual ACM international workshop on Web information and data management*, pp. 67–74, 2005.

[5] S. Deerwester, S. T. Dumais, G. W. Furnas, T. K. Landauer and R. Harshman. Indexing by latent semantic analysis. *Journal of the American Society for Information Science*, 41(6): 391–407, 1990.

[6] A. P. Dempster, N. M. Laird and D. B. Rubin. Maximum likelihood from incomplete data via the em algorithm. *Journal of the Royal Statistical Society, Series B*, 39(1):1–38, 1977.

[7] F. R. Dwyer. Customer lifetime valuation to support marketing decision making. *Journal of Direct Marketing*, 11(4): 6–13, 1997.

[8] J. L. Herlocker, J. A. Konstan, A. Borchers and J. Riedl. An algorithmic framework for performing collaborative filtering. In *SIGIR '99: Proceedings of the 22nd annual international ACM SIGIR conference on Research and development in information retrieval*, pp. 230–237, 1999.

[9] J. L. Herlocker, J. A. Konstan, L. G. Terveen and J. T. Riedl. Evaluating collaborative filtering recommender systems. *ACM Trans. Inf. Syst.*, 22: 5–53, January 2004.

[10] T. Hofmann. Probabilistic latent semantic indexing. *In: Proceedings of the 22nd an-nual international ACM SIGIR conference on Research and development in information retrieval*, pp. 50–57, 1999.

[11] T. Hofmann and J. Puzicha. Latent class models for collaborative filtering. *In: IJCAI*, pp. 688–693, 1999.

[12] S. K. Lam, D. Frankowski and J. Riedl. Do you trust your recommendations? an exploration of security and privacy issues in recommender systems. *In: ETRICS*, pp. 14–29, 2006.

[13] S. K. Lam and J. Riedl. Shilling recommender systems for fun and profit. *In: Proceedings of the 13th international conference on World Wide Web*, pp. 393–402, 2004.

[14] R. J. Mooney and L. Roy. Content-based book recommending using learning for text categorization. *In: Proceedings of the fifth ACM conference on Digital libraries*, pp. 195–204, 2000.

[15] M. Papagelis and D. Plexousakis. Qualitative analysis of user-based and item-based prediction algorithms for recommendation agents. *Eng. Appl. Artif. Intell.*, 18: 781–789, October 2005.

[16] S. Rosset, E. Neumann, U. Eick, N. Vatnik and Y. Idan. Customer lifetime value modeling and its use for customer retention planning. *In: Proceedings of the eighth ACM SIGKDD international conference on Knowledge discovery and data mining*, pp. 332–340, 2002.

[17] B. Sarwar, G. Karypis, J. Konstan and J. Reidl. Item-based collaborative filtering recommendation algorithms. *In: Proceedings of the 10th international conference on World Wide Web*, pp. 285–295, 2001.

[18] B. M. Sarwar, J. A. Konstan, A. Borchers, J. Herlocker, B. Miller and J. Riedl. Using filtering agents to improve prediction quality in the grouplens research collaborative filtering system. *In: Proceedings of the 1998 ACM conference on Computer supported cooperative work*, pp. 345–354, 1998.

[19] D. C. Schmittlein, D. G. Morrison and R. Colombo. Counting your customers: who are they and what will they do next? *Manage. Sci.*, 33: 1–24, January 1987.

[20] X. Su and T. M. Khoshgoftaar. A survey of collaborative filtering techniques. *Adv. in Artif. Intell.*, 4:2–4:2, January 2009.

[21] J. Sun, D. Tao and C. Faloutsos. Beyond streams and graphs: dynamic tensor analysis. *In: Proceedings of the 12th ACM SIGKDD international conference on Knowledge discovery and data mining*, pp. 374–383, 2006.

[22] Y. Yang and B. Padmanabhan. Evaluation of Online Personalization Systems: A Survey of Evaluation Schemes and A Knowledge-Based Approach. *Journal of Electronic Commerce Research*, 6(2): 112–122, 2005.

[23] K. Yu, X. Xu, J. Tao, M. Ester and H.-P. Kriegel. Instance Selection Techniques for Memory-Based Collaborative Filtering. *In: Proceedings of the 2nd SIAM International Conference on Data Mining*, 2002.

CHAPTER 12

Social Tagging Systems

In this chapter, we present the literature review on works on tag-based systems. First, we introduce the background of this chapter. The purpose of literature review is to gain insight in research that has been done already; in turn this process enables us to identify useful ideas, unsolved issues and shortcomings in the current methods.

Accordingly, this chapter is structured as follows: in Section 12.1 we explore the literature on the basic concept of data mining and information retrieval which links this chapter to the whole book; in Section 12.2 the related work in recommender systems in details, we review the recommendation algorithms, and discuss the tag-based recommender system; subsequently we review the clustering algorithms which helps to improve the recommendation in Section 12.3. Following that in Section 12.4 we discuss the Clustering algorithms in Tag-Based Recommender Systems in details. Finally, a summary is given in Section 12.5.

12.1 Data Mining and Information Retrieval

Data mining is one of the popular research areas which has a long processing of research period. Data mining can take the evolutionary process and analyze data from various perspectives, and then summarize the useful information for the users [53]. We want to utilize the techniques and algorithms from data mining to process the data in the various fields to find the correlations with the different attributes.

Information Retrieval is a broad but full of challenge part of research areas. It also has a long history, as early as 1968, when Lancaster [33] gave the perfectly straightforward idea. It mainly focuses on providing useful and helpful information to users by easy access.

Information retrieval (IR) should provide the useful and interesting information according to the users' need with easy access from the dataset. And it also helps to represent, store and reorganize the information items [3].

When a user enters a query into an information retrieval system, the system will compute a numeric score based on the similarity between each object and such query, and then rank the objects to generate the ranking list, so the value will present how well each object in the database matches the query [34].

There are various IR systems based on the different user queries, such as text documents, images, [18] audio, [14] mind maps [4] or videos. We involved different approaches to improve the calculation of ranking scores with different degrees of relevancies; they can obtain the top ranked objects to generate better recommendations.

In [8], an information retrieval system is composed in three parts:

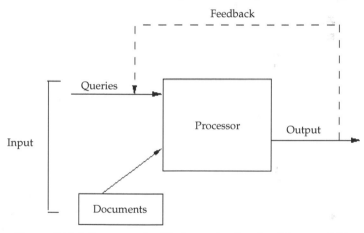

Figure 12.1.1: Framework of Information Retrieval System ([8])

The representation of the documents in the dataset and the query from user will initialize as the input part in the beginning.

And then, some of the techniques and algorithms will involve to structure information in an appropriate way. It will also involve performing the actual retrieval function. The output part is usually a set of document lists.

12.2 Recommender Systems

The search engines based on the data mining technique can help user to obtain sufficient information resources; however, they have to filter the bulky information resources themselves to get proper information. So how to deal with the problems of ambiguity and redundancy represents the urgent and challenging needs. Thus the recommender systems came into birth according to the basic needs for the different users.

The recommender system provides a list of recommended items to the user by calculating the similarity between the collected data and the documents in the dataset. It can also help users to discover the useful items they might not have found by themselves.

The first overview of recommender systems was from an intelligent agent's perspective which was provided by Montaner [44]. Herlocker et al. [23] surveyed the evaluation techniques for recommender systems.

The recommender systems improve the performance of the search engines to index the non-traditional data. Tag has been widely used as an additional attribute in recommender systems. Tag is a kind of metadata which helps to describe an item and allows it to be found again by browsing or searching. It is assigned by the individual user to the web resource which can represent the user's personal opinion expression [7]. The websites like Del.icio.us, Last.fm, and Flickr are the masterpieces of Web 2.0's applications. They allow users to express their own preferences on the original resources with freely annotated words. So how to use the social tagging data for better recommendation in appropriate way becomes an active research topic recently.

Tags are used as an additional feature to re-model users or resources over the tag vector space, and the annotation attribute can improve the personalized recommendation. Different users can annotate the various tags on the same resource, and the same tags also can be annotated by different users as well [40].

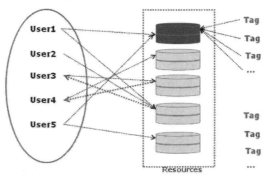

Figure 12.2.1: Relationship of Users, Tags, Resources in Tagging System

Before we start digging into tag based recommender systems we will shortly discuss developments in the recommender systems area in general.

12.2.1 Recommendation Algorithms

In the previous subsection, we introduced the primary processing of recommendation, and we will review the current recommendation algorithms in this section. Overall, the recommendation algorithms are utilized to recommend items which users are searching for currently, or predict the items that they have not considered yet. Adomavicius and Tuzhilin formulate the recommendation problem as follows [15]:

Let $U = \{u_1, \cdots u_m\}$ be a set of users, and let $I = \{i_1, \cdots i_m\}$ be a set of items. Let $U \times I \rightarrow R$, where R is a totally ordered set, and the $g(u_m, i_n)$ measures the similarity between item in to user u_m. Then, we want to recommend unknown item $i^{max,u} \in I$ to the user $u \in U$, which maximize the function g:

$$\forall u \in U, i^{max,u} = \arg\max_{i \in I} g(u,i)$$

The basic concept of recommendation algorithms is shown as above, which provides the fundamental background to the research. Below we will introduce the Collaborative Filtering techniques and Content-based techniques which are two basic types of recommendation methods.

12.2.1.1 Collaborative Filtering Recommendation

Collaborative Filtering (CF) is a mellow technique which has been widely used in the recommender systems. It processes for filtering information or patterns using techniques involving collaboration among multiple conditions [24]. The collaborative filtering typically focuses on user data from very large data sets. Generally speaking, it is based on the user's historical behavior that means when the user is interested in an item in the past; it will be the same in future. If another user who is interested in the same item, the system will define them have one of the common options. Then the system will provide recommendation according to the same preferences of them.

It can produce personal recommendations by computing the similarity between the user's preference and other related people.

Figure 12.2.2: Principle of Collaborative Filtering Recommendation

The basic mechanism behind collaborative filtering systems is the following:

- Collect a large group of people's preferences;
- Analyze the similarity among a subgroup of people;
- Select the people who has the similar preferences as the person who seeks advice;
- Calculate the average preferences score for that subgroup people;
- Generate the recommended items to the user based on preference function.

The collaborative filtering has several mechanisms as below:

12.2.1.1.1 Memory-based It utilizes users' rating data to compute similarity between users or items. In principle, there are neighborhood-based collaborative filtering and item-based or user-based top-N recommendations [62].

The neighborhood-based algorithm calculates the similarity between two users or items. It predicts the average preferences score for all of the rating items. We involve the mechanisms of cosine similarity on vector space for our research work. The approach we implemented is as below:

Firstly the algorithm calculates the similarity value on the vector model, secondly it collects the k most similar users by using top-N recommendation algorithm, and then, it aggregates the user item matrices corresponding to the identified k most similar users; finally it can identify the set of items to be recommended.

In addition, another popular method used to find the similar users is called the Locality Sensitive Hashing. It implements the nearest neighbor mechanism in linear time.

The advantages of this approach are: The result is convenient to explain; the implementation is easy to create and use; when the system has the new data, it can be updated easily and incrementally; the content of the recommended items do not need to be considered.

However, there are several disadvantages of this approach: First, it depends on the users' rating histories, so it has the limitation called "cold-start", meaning that systems can only generate the recommendation when there are enough user data. Second, it has the poor prediction with the large dataset especially when data get sparse or the number of similar users is small. Third, it cannot generate the recommendation for the new users or the users without rating histories.

12.2.1.1.2 Model-based Models are developed by using data mining and machine learning algorithms to find patterns based on training data. There are many algorithms such as Bayesian Networks, clustering models, Markov decision process based models, and so on. The classification and clustering techniques help the models to identify the user with the different parameters. The number of the parameters can be changed by different types according to principal component analysis [63].

The advantages of this approach are: It has the better preference on the sparsity, so it is more suitable for large data sets on the prediction performance. It provides the recommendation with more intuitive rationale.

The disadvantages of this approach are: It is difficult to explain the predictions for some of the models. Modeling process is more complex. It is difficult to gain both well prediction performance and scalability. Some of the useful information would be lost by reducing models.

12.2.1.1.3 Hybrid It is based on the combination of the memory-based and the model-based CF algorithms. Such technique improves the scalability of model-based approach and the accuracy of memory-based approach; therefore, it performs more effectively than both of them. In addition, it solves the problem of data sparsity. However, it increases the complexity to implement [64].

12.2.1.2 Content-based Recommendation

Content-based filtering recommendation is based on the content similarity of the items. It aims on recommending items based on the idea that if a user liked an item in the past which had been recorded by the system, he/she might probably like other similar items in the future. The system collected attributes for the items from the previous information, and then provided the recommended items. The recommendation decision is made by comparing the candidate items with the previously rated item. The best-matching items are recommended to the users [65].

In content-based recommendation approaches, the function $g(u_m, i_n)$ is formulated as:

$$g(u_m, i_n) = sim(ContentBasedUserProfile(u_m), ContentBasedItemProfile(I_n))$$

Where $ContentBasedUserProfile(u_m)$ is composed by content-based user preferences of a user $u \in U$, and the $ContentBasedItemProfile(I_n)$ is the set of content features characterizing item $i \in I$.

Basically, the above method characterizes items within the system by item profile. Then, different characteristics of item are expressed as a score vector. Finally, the system forms a content-based profile of users based on score vector.

Overall, the recommendation scores denote the importance of each characteristic to the user. The recommendation scores can be calculated from individually rated content vectors.

Alternatively, calculating the similarity between the attributes that the user preferred and those are not preferred is also a method of generating the recommendation score. The scores can then be used to estimate the probability of a specific part of the attributes that is potentially preferred by the same user. Some other methods to calculate the recommendation scores estimate the preferences of the users over the items by utilizing machine learning techniques [66].

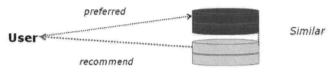

Figure 12.2.3: Principle of Content-based Recommendation

Web pages or other kinds of documents can be clustered into the same group by the same characteristics; the system stores the relationship among users, tags and documents. When the users have the similar preferred document, the system can recommend other documents to him which can be collected from the other users with the same experience hobbies.

The advantages of this approach are: It is easy to establish the content-based user profile based on the weighted vector of item attributes, without a need of other users' data; it can explain the recommendation by listing content-feature items; there is no "First-Rater Problem" for the new item.

The disadvantages of this approach are: It excessively depends on the particular user relevance; it provides recommendation relying on all content information, whereas there is only a very shallow analysis of content that can be supplied, and the content must be encoded as meaningful features; it also has the problem of "over-specialization", that means, the system can only recommend users with the similar items that they have already preferred, to the user according to the highest score.

12.2.2 Tag-Based Recommender Systems

Tagging system has some advantages in [48] as: Low cognitive cost and entry barriers; immediate feedback and communication; individual needs and information of organization.

The simple tagging system allows any web user to annotate the free words on their favorite web resources rather than the predefined vocabulary. Users can communicate with each other implicitly by the tag suggestions to describe resources on the web. Therefore, the tagging system provides a convenient way for users to organize their favorite web resources. In addition, due to the development of the system, the user can find other people who are interested in similar projects. Consensus around stable distributions and shared vocabularies emerge [21], even in the absence of a centrally controlled vocabulary.

12.2.2.1 Folksonomy

When users want to annotate web documents for better organization and use the relevant information to retrieve their needed resources later, they often comment such information with free-text terms. Tagging is a new way of defining characteristics of data in Web 2.0 services. The tags help users to collectively classify and find information and they also represent the preference and interests of users. Similarly, each tagged document also expresses the correlation and the attribute of the document. A kind of data structure can be established based on the tagging annotation.

Hotho et al. [26] combined users, tags and resources in a data model called *folksonomy*. It is a system which classifies and interprets contents. It is the derivative of the method of collaboratively creating and organizing tags.

Folksonomy is a three-dimensional data model of social tagging behaviors of users on various documents. It reveals the mutual relationships between these three-fold entities, i.e. user, document and tag. A folksonomy F according to [26] is a tuple $F = (U, T, D, A)$, where U is a set of users, T is a set of tags, D is a set of web documents, and $A \subseteq U \times T \times D$ is a set of annotations. The activity in folksonomy is $t_{ijk} \subseteq \{(u_i, d_j, t_k) : u_i \in U, d_j \in D, t_k \in T\}$, where $U = \{U_1, U_2, \cdots, U_M\}$ is the set of users, $D = \{D_1, D_2, \cdots, D_N\}$ is the set of documents, and $T = \{T_1, T_2, \cdots, T_K\}$ is the set of tags. $t_{ijk} = 1$ if there is an annotation (u_i, d_j, t_k); otherwise $t_{ijk} = 0$.

Therefore a social tagging system can be viewed as a tripartite hypergraph [43] with users, tags and resources represented as nodes and the annotations represented as hyper-edges connecting users, resources and tags. There are some social applications which are based on the folksonomy such as social bookmarking and movies annotation.

In this section, the preliminary approach for recommender system is based on the folksonomy model, which helps us to obtain the tagging information, and generate the user profile, document profile and group profiling.

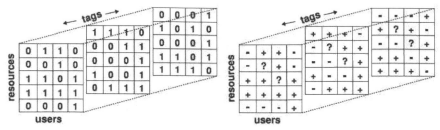

Figure 12.2.4: Relationship of Users, Tags, Resources in Folksonomy

The advantage of the folksonomy is to combine the three-dimensional data into one data model; each two parts can represent the related information, furthermore it is much more convenient for analyzing the users' behaviors and the documents' attributes in the folksonomy model.

12.2.2.2 Standard Recommendation Model in Social Tagging System

Standard social tagging systems may vary in the ways of their ability of handling recommendation. In this subsection, we focus our discussion on the folksonomy model, which is derived from the information retrieval principle. In folksonomy model, each user can be represented in the tag set vector. Tag frequency represents the popularity of different tags. We use the tag frequency as [25], $TF = |a = \langle u, r, t \rangle \in A : u \in U, r \in R, t \in T|$, to calculate the weight of the vector, which means, if a user u, has an annotation A, and he assigns a tag t, on a resource r, such behavior will be assigned as "1" in the tagging matrix; otherwise "0", so the user can be represented as $u = \langle utf(t_1), utf(t_2), \cdots, utf(t_{|T|}) \rangle$, Likewise each resource, r, can be modelled as $r = \langle rtf(t_1), rtf(t_2), \cdots, rtf(t_{|T|}) \rangle$.

There are various similarity measures such as the Jaccard Coefficient, Pearson Correlation or Cosine similarity to calculate the similarity scores, and there are different approaches based on the user vector or resource vector. The system provides top-N items as the recommendation list according to the ranked similarity values.

There are several other recommendation algorithms proposed to generate the recommendation list, such as FolkRank algorithm, LocalRank algorithm, and so on. The FolkRank is enlightened by the [67], the basic idea for FolkRank is that if an important user annotated a resource by an important tag, then, such resource would be important, the recommendation is based on calculating the importance weight [26]. Kubatz et al. [68] improved the FolkRank by utilizing a neighborhood-based tag recommendation algorithm called LocalRank, focuses on the relevant ones only, and the recommendation accuracy is on a par with or slightly better than FolkRank.

12.3 Clustering Algorithms in Recommendation

The traditional recommendation algorithms such as collaborative filtering approach, content-based filtering approach, and so on, are too much reliant on users' data and such data generally has the problem of sparseness. When collecting the user profiles by the approaches above, the sparse data would exacerbate the computational complexity and reduce the precision of recommendation. So we consider involving the clustering algorithms to reduce the dimensions of users and documents data. With the help of clustering algorithms, both recommendation performance and results can be improved.

Clustering algorithms refer to algorithms which are trying to find hidden structures in unlabeled data. The clustering algorithms are used to estimate, summarize and explain the main characteristic of the data. There are many cluster methods which are based on data mining [30].

We will introduce the *K*-means, hierarchical clustering and density based clustering in the following sections.

12.3.1 K-means Algorithm

The *K*-means clustering algorithm assigns the objects into *k* number of clusters based on the various factors; it is a top-down algorithm. *k* is a positive integer number and specified apriority by users. The processing is finished by minimizing the sum of squares of distances between data and the corresponding cluster centroid [52].

The basic idea behind *K*-means is as follows: In the beginning the number of clusters *k* is determined. Then the algorithm assumes the centroids or centers of these *k* clusters. These centroids can be randomly selected or designed deliberately. One special case is when the number of objects is less than the number of clusters. If such case exists, each object is set as the centroid of the individual cluster and assigned a cluster number. If the number of objects is bigger than the number of clusters, the algorithm calculates the distance (i.e., Euclidean distance) between each object and all of the centroids to obtain the minimum distance. When the process starts, the centroid location is unknown, so algorithm updates centroid location according to the processed information, such as the minimum distance between the objects and the new centroids. When all of the objects are assigned to the *k* clusters, the centroids have finished updating. Such above process repeats until there are no longer large changes for assigning the objects into the clusters, or centroids do not change in successive iterations. So the iteration convergence can be proved mathematically [19].

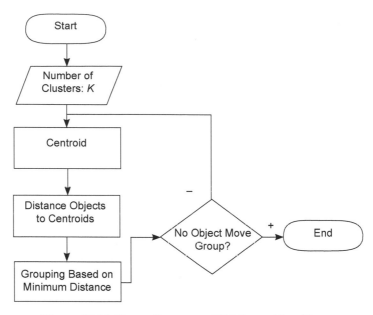

Figure 12.3.1: Frame Structure of K-Means Algorithm

Description: Given a set of observations $(x_1, x_2, \cdots, x_n)$, where each observation is a d-dimensional real vector, k-means clustering aims to partition the n observations into k sets $(k \leq n)$, $S = \{S_1, S_2, \cdots, S_k\}$ so as to minimize the Within-Cluster Sum of Squares (WCSS), where μ_i is the mean of points in S_i.

$$\arg\min_{S} \sum_{i=1}^{k} \sum_{x_j \in S_i} \| x_j - \mu_i \|^2$$

The advantages of the K-means algorithm are: The low time consumption and the fast processing speed on the condition of value k is small; the compacted clusters production performance is satisfactory; the clusters do not overlap since they are in non-hierarchical structure.

There are some disadvantages of K-means algorithm: The algorithm is not able to calculate the applicable number of clusters automatically; the user has to assign the value k as an input to the algorithm in advance. Simultaneously, the specific number of clusters restricts the prediction of what the real k should be. Various initial partitions lead to different number of clusters, and the results for different composition of clusters can be distinct in some of the experiments.

There are extensive related research works on it. The author in [27] theorized that *K*-means was a classical heuristic clustering algorithm. Due to the sensitivity problem of *K*-means, some modified approaches have been proposed in the literature. Fast Global *K*-means [51] (FGK means), for example, is an incremental approach of clustering that dynamically adds one cluster centre at a time through a deterministic global search procedure consisting of *D* executions of the *K*-means algorithm with different suitable initial positions. Zhu et al. presented a new clustering strategy, which can produce much lower $Q(C)$ value than affinity propagation (AP) by initializing *K*-means clustering with cluster centre produced by AP [45]. In [42], the authors were motivated theoretically and experimentally by a use of a deterministic divisive hierarchical method and use of PCA-part (Principal Component Analysis Partitioning) as the initialization of *K*-means. In order to overcome the sensitivity problem of heuristic clustering algorithm, Han et al. proposed CLARANS based on the random restart local search method [4]. VSH [9] used the iteratively modifying cluster centre method to deal with initiation problem. More modified methods addressing the initialization sensitivity problem of clustering algorithm are referred to [20, 36, 59] .

12.3.2 Hierarchical Clustering

The *K*-means algorithm has the limitation of choosing the specific number of clusters, and it has the problem of non-determinism. It returns the clusters in an unstructured set. As a result of such limitations, if we require hierarchy structure, we need to involve the hierarchical clustering.

Hierarchical clustering constructs a hierarchy of clusters that can be illustrated in a tree structure as a dendrogram. Each node in the tree structure, including the root, represents the relationship between parents and children, so it is able to explore different levels of clustering granularity [19]. Hierarchical clustering algorithms are either top-down or bottom-up, the bottom-up algorithms treat each file as a separate cluster in the beginning and then begin to merge, until all cluster clusters have been merged into a single cluster, such cluster contains all the files.

The bottom-up hierarchical clustering is called hierarchical agglomerative clustering.

The top-down clustering requires a method for dividing a cluster. It splits clusters recursively until the individual documents are reached [69]

The advantages of the hierarchical clustering are [5, 19]: It has a high flexibility with respect to the level of granularity; it is easy to deal with any form of similarity metric or the distance; it does not require pre-assignment of the number of clusters, and therefore has high applicability.

The disadvantages of the hierarchical clustering are summarized as [70]: The termination judgment conditions and the interpretation of the hierarchy are complex; if an incorrect assignment exists, most hierarchical algorithms do not rebuild intermediate clusters; the single pass of analysis and local decisions are the influencing factor of the clusters.

12.3.3 Spectral Clustering

The spectral clustering combines some of the benefits of the two aforementioned approaches. It refers to a class of techniques which rely on the eigenvalues of the adjacency similarity matrix; it can partition all of the elements into disjoint clusters, the elements that have high similarity will end up in the same cluster. Elements within one cluster have low similarity with other clusters' elements. The spectral clustering is based on the graph partition. It maps the original inherent relationships onto a new spectral space. The whole items are simultaneously partitioned into disjoint clusters with minimum cut optimization. Spectral clustering techniques make use of the spectrum of the similarity matrix of the data to perform dimensionality reduction for clustering in fewer dimensions [71].

The original formula for the spectral clustering is:

$$L = I - D^{-1/2}WD^{-1/2}$$

where W is the corresponding similarity matrix, and D is the diagonal matrix, $D_{ii} = \sum_j S_{ij}$.

According to the spectral graph theory in [13], the k singular vectors of the reformed matrix $RM_{User} = D^{-1/2}SM_{User}D^{-1/2}$ present a best approximation to the projection of user-tag vectors on the new spectral space.

Compared to those clustering algorithms above, spectral clustering algorithm has many fundamental advantages: It is very simple to implement; it performs well with no local minima, so it could be solved efficiently by standard linear algebra methods; it also can keep the shapes and densities in the cluster invariantly; the performance of obtained result is better.

The disadvantages of the spectral clustering are summarized as: The high time complexity and space complexity lead the processing inefficient. In some cases, the clustering processing is unstable.

Example of data set

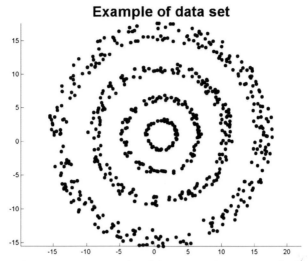

Figure 12.3.2: Example of the Spectral Clustering [37]

Clustering results :

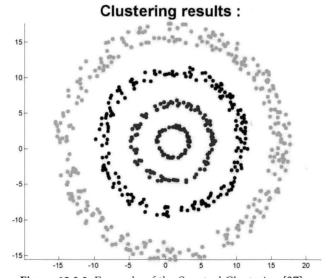

Figure 12.3.3: Example of the Spectral Clustering [37]

12.3.4 Quality of Clusters and Modularity Method

There are various categories of methods to measure the quality of clusters, such as "Compactness", a measure of similarity of objects within an individual cluster to the other objects outside the cluster; or the "Isolation", a measure of separation among the objects outside the cluster [54]. In the research, we combine such attributes together, so as to utilize the modularity method to evaluate the clustering algorithms. It is one of the quantitative measures for the "goodness" of the clusters discovered.

The modularity value is computed by the differences between the actual number of edges within a cluster and the expected number of such edges. The high value of the modularity shows the good divisions; that means, the nodes within the same cluster have the concentrated connections but only sparse connections between different clusters. It helps to evaluate the quality of the cluster; here "quality of cluster" consists of two criteria, i.e., the number of clusters and the similarity of each cluster [32].

Consider a particular division of a network into k clusters. We can define a $k \times k$ symmetric matrix SM whose element sm_{ij} is the fraction of all edges in the network that link vertices in cluster p to vertices in cluster q. Take two clusters C_p and C_q randomly, the similarity smC_{pq} between them can be defined as

$$smC_{pq} = \frac{\sum\limits_{c_p \in C_p} \sum\limits_{c_p \in C_q} c_{pq}}{\sum\limits_{c_q \in C} \sum\limits_{c_p \in C} c_{pq}}, p,q = 1,2\cdots m$$

where c_{pq} is the element in the similarity matrix for the whole objects. When $p=q$, the smC_{pq} is the similarity between the elements inside the clusters, while $p \neq q$, the smC_{pq} is the similarity between the cluster C_p and the cluster C_q. So the condition of a high quality cluster is $\max(\sum\limits_p smC_{pp})$ and $\min(\sum\limits_{p,q} smc_{pq}), p \neq q, p, q = 1, 2, \cdots m$.

Summing over all pairs of vertices in the same group, the modularity, denoted Q, is given by:

$$Q = \sum\limits_{q=1}^{m} [smc_{pp} - (\sum\limits_{q=1}^{m} smc_{pq})^2] = \text{Tr}SM - \| SM^2 \|$$

where the value m is the amount of clusters. The trace of this matrix $\text{Tr}SM$ $\sum\limits_{p=1}^{m} smC_{pp}$ gives the fraction of edges in the network that connect vertices

in the same cluster, and a good division into clusters should have a high value of it. If we place all vertices in a single cluster, the value of $\text{Tr}SM$ would get the maximal value of 1 because there is no information about cluster structure at all.

This quantity measures the fraction of the edges in the network that connect vertices of the same type minus the expected value of the same quantity in a network with the same cluster divisions. Utilize the value Q to evaluate the clusters [4]: Values approaching $Q=1$, which is the maximum, indicate that the whole network has a strong cluster structure. In practice, values for such networks typically fall in the range from about 0 to 1. The higher value of Q, the better quality for the cluster the C_p and C_q is, so that we can get the optimal number of clusters.

12.3.5 K-Nearest-Neighboring

In *KNN* algorithm, the object is classified by the neighbors who have been separated into several groups, and the object is assigned into the class which has the most common neighbors amongst its k nearest majority influence neighbors. The *KNN* algorithm is sensitive to the local data structure. The training data of the algorithm is the neighbors who are taken from a set of objects with the correct classification. In order to identify neighbors, the objects are represented in the multidimensional feature space vectors [22].

k is a positive integer, it is typically small. Take an example in Fig. 12.3.4, if $k=1$, then the object is simply assigned the class of its nearest neighbor. In binary (two class) classification problems, it is helpful to choose k to be an odd number as this avoids difficulties with tied votes [12, 49].

The test sample red triangles should be classified either to the first class of green circle or to the second class of blue star. If $k = 3$ it should be classified to the first class because there are 2 green circles and only 1 blue star inside the inner circle. If $k = 5$ it should be classified to second class since there are 3 stars and only 2 circles inside the outer circle.

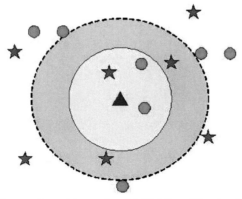

Figure 12.3.4: Example of *KNN* Classification [1]

The advantages of the *KNN* algorithm are: Such algorithm is easy to implement; it has a strong applicability, although the prediction accuracy can be quickly degraded when the number of attributes grows.

The disadvantages of the *KNN* algorithm are: It needs to compare the test item with all of the items in the training set, so the time complexity is higher than the linear classifier when it makes the predictions; and its performance depends too much upon the similarity and the k value.

KNN algorithm adaptation methods have been widely used in the tag classification. Cheng et al. combine the *KNN* method and logistic regression to exploit the multiple dependence [11]. Zhang et al. propose *ML-KNN*, a lazy method that firstly finds k neighbors of the test instance, and then gives the predicted label set by maximizing each labels posterior [57].

In this chapter, we aim on the major problem of most social tagging systems resulting from the severe difficulty of ambiguity, redundancy and less semantic nature of tags. We employ the *KNN* algorithm to establish the structure for potential relationship information of the tags neighbors. Then we combine the *KNN* graph with the clustering algorithm to filter the redundant tags neighbors for improving the recommendation performance.

12.4 Clustering Algorithms in Tag-Based Recommender Systems

As tags are of syntactic nature, in a free style and do not reflect sufficient semantics, the problems of redundancy, ambiguity and less semantics of tags are often incurred in all kinds of social tagging systems [47]. For example, for one resource, different users will use their own words to describe their feeling of likeness, such as "favourite, preference, like" or even the plural form of "favourites"; and another obstacle is that not all users are willing to annotate the tags, resulting in the severe problem of sparseness.

In order to deal with these difficulties, clustering methods have been introduced recently into social tagging systems to find meaningful information conveyed by tag aggregates. In past years, many studies have been carried out on tags clustering. Gemmell et al [16, 50] demonstrated how tag clusters serving as coherent topics can aid in the social recommendation of search and navigation. The aim of tag clustering is to reveal the coherence of tags from the perspective of how resources are annotated and how users annotate in the tagging behaviors. Undoubtedly, the tag cluster form is able to deliver user tagging interest or resource topic information in a more concise and semantic way. It handles to some extent the problems of tag sparseness and redundancy, in turn, facilitating the tag-based recommender systems. Thus this demand mainly motivates the research of tag clustering in social annotation systems. In general, the tag clustering algorithm could be described as: (1) Define a similarity measure of tags and construct a tag similarity matrix; (2) Execute a traditional clustering algorithm such as *K*-Means [16, 50], or Hierarchical Agglomerative Clustering on this similarity matrix to generate the clustering results; (3) abstract the meaningful information from each cluster and do recommendation [59].

Martin [38] et al. propose to reduce tag space by exploiting clustering techniques so that the quality of the recommendations and execution time are improved and memory requirements are decreased. The clustering is motivated by the fact that many tags in a tag space are semantically similar thus the tags can be grouped.

Astrain et al. firstly combines a syntactic similarity measure based in a fuzzy automaton with ε-moves and a cosine relatedness measure, and then design a clustering algorithm for tags to find out the short length tags [2]. In general, tags lack organizational structure limiting their utility for navigation. Simpson proposes a hierarchical divisive clustering algorithm to release these influence of the inherent drawback of tag data [4]. In [6], an approach that monitors users' activity in a tagging system and dynamically quantifies associations among tags is presented and the associations are then used to create tags clusters. Zhou et al. propose a novel method to compute the similarity between tag sets and use it as the distance measure to cluster web documents into groups [58].

In [10], clusters of resources are shown to improve recommendation by categorizing the resources into topic domains. A framework named Semantic Tag Clustering Search, which is able to cope with the syntactic and semantic tag variations, is proposed in [55]. And in [39] topic relevant partitions are created by clustering resources rather than tags. By clustering resources, it improves recommendations by distinguishing between alternative meanings of query. While P. Lehwark et al. use Emergent-Self-Organizing Maps (ESOM) and U-Map techniques to visualize and cluster tagged data and discover emergent structures in collections of music [35]. State-of-the-art methods suffice for simple search, but they often fail to handle more complicated or noisy web page structures due to the key limitations. Miao et al. propose a new method for record extraction that captures a list of objects in a more robust way based on a holistic analysis of a web page [41]. In [17], a co-clustering approach is employed, which exploits joint groups of related tags and social data resources, in which both social and semantic aspects of tags are considered simultaneously. The common characteristic of aforementioned tagging clustering algorithm is that they use K-Means or hierarchical clustering algorithms on tag dataset to find out the similar tag groups. In [46], however, the authors introduce Folks Engine, a parametric searching engine for folksonomies allowing specifying any tag clustering algorithm. In a similar way, Jiang et al., make use of the concept of ensemble clustering to find out a consensus tag clustering results of a given topic and propose tag groups with better quality [29]. The efficient way which improves tag clustering result is to use the common parts of several tag clustering results. Approximate Backbone, the intersection of different solutions of a dataset, is often used to investigate the characteristic of a dataset [61, 28]. Zong et al. use approximate backbone to deal with the initialization problem of heuristic clustering algorithm [60].

Alexandros et al. [31] focused on the complexity of social tagging data. They developed a data-modeling scheme and a tag-aware spectral clustering procedure. They used tensors to store the multi-graph structures and capture the personalized aspects of similarity. They present the similarity-based clustering of tagged items, and capture and exploit the multiple values of similarity reflected in the tags assigned to the same item by different users. Also they extend spectral clustering by capturing multiple values of similarity between any two items. The authors above focus on calculating similarity approach to improve the spectral clustering, however, how to evaluate the quality of clusters is not mentioned.

In this section, we investigate the clustering algorithms used in social tagging systems. With the help of clustering algorithms, we can obtain the potential relationship information among the different users and various resources, and clustering also reduces the dimensionality in calculation. The clusters can reduce the time complexity in recommendation processing. In a word, the clustering algorithms help to enhance the tag expression quality and improve the recommendation in social tagging systems.

12.5 Chapter Summary

In this chapter, we have reviewed the basic concept of data mining and information retrieval techniques used in recommender systems, such as clustering and K-Nearest-Neighboring. This chapter has also discussed the data mining problems existed in the social tagging system, raised some of the current techniques, and investigated advantages and disadvantages of such approaches, which provide a guideline for dealing with recommendation problems and improving the performance of recommendation.

Reference

[1] A. Ajanki. Example of k-nearest neighbour classification, 2007.
[2] C. A. e. a. Astrain J. J. and Echarte F. A tag clustering method to deal with syntactic variations on collaborative social networks, 2009.
[3] R. A. Baeza-Yates and B. Ribeiro-Neto. *Modern Information Retrieval*. Addison-Wesley, New York, 1999.
[4] J. Beel, B. Gipp and J.-O. Stiller. Information retrieval on mind maps—what could it be good for?, 2009.
[5] P. Berkhin. Survey of clustering data mining techniques. Technical report, Accrue Software, 2002.
[6] V. E. Boratto L. and Carta S. Ratc: A robust automated tag clustering technique, 2009.
[7] S. P. Borovac and Mislav. Expert vs. novices dimensions of tagging behaviour in an educational setting. *Bilgi Dinyasi*, (13 (1)): 1–16, 2012.
[8] P. M. F. C. F. C. J. van Rijsergen. *Information Retrieval*.

[9] J. G. Cao F. Y. and Liang J. Y. An initialization method for the *k*-means algorithm using neighborhood model. *Computers and Mathematics with applications*, 58(3) (pp. 474–483), 2009.

[10] D. S. Chen, H. Bringing order to the web: Automatically categorizing search results, 2000.

[11] W. Cheng and E. Hullermeier. Combining instance-based learning and logistic regression for multilabel classification. *Machine Learning*, 76 (Number 2-3): 211, p. 225, 2009.

[12] B. V. Dasarathy. *Nearest Neighbor (NN) Norms: NN Pattern Classification Techniques.* 1991.

[13] I. S. Dhillon. Co-clustering documents and words using bipartite spectral graph partitioning, 2001.

[14] J. Foote. An overview of audio information retrieval. *Multimedia Systems*, 1999.

[15] A. G. Adomavicius, Tuzhilin. Toward the next generation of recommender systems: A survey and possible extensions. *IEEE Transactions on Knowledge & Data Engineering*, 17(6): 734–749, 2005.

[16] J. Gemmell, A. Shepitsen, M. Mobasher and R. Burke. Personalization in folksonomies based on tag clustering, July 2008.

[17] K. V. V. A. K. Y. Giannakidou, E. Co-clustering tags and social data sources, 2008.

[18] A. A. Goodrum. Image information retrieval: An overview of current research. *Informing Science*, 3(2), 2000.

[19] L. L. Guandong Xu and Yanchun Zhang. *Web mining and social networking: techniques and applications.* Web information systems engineering and Internet technologies. New York: Springer 2011.

[20] M. K. Hariz, S. B. and Elouedi Z. Selection initial modes for belief k-modes method. international journal of applied science. *Engineering and Tchnology*, 20084(4): 233–242, 2008.

[21] H. S. Harry Halpin and Valentin Robu. The complex dynamics of collaborative tagging, 2007.

[22] A. R. E. Hector Franco-Lopez and M. E. Bauer. Estimation and mapping of forest stand density, volume, and cover type using the k-nearest neighbors method. *Remote Sensing of Environment*, pp. 251–274, September 2001.

[23] J. L. Herlocker, J. A. Konstan, L. G. Terveen and J. T. Riedl. Evaluating collaborative filtering recommender systems. *ACM Trans, Inf. Syst.* 22 (1), January 2004.

[24] L. T. Hill and Will. Beyond recommender systems: Helping people help each other. *HCI in the New Millennium*, Addison-Wesley pp. 487–509, 2001.

[25] A. Hotho, R. Jschke, C. Schmitz and G. Stumme. Folkrank: A ranking algorithm for folksonomies. *In Proc. FGIR 2006*, 2006.

[26] A. Hotho, R. Jschke, C. Schmitz and G. Stumme. Information retrieval in folksonomies: Search and ranking, June 2006.

[27] A. Jain and R. Dubes. *Algorithms for clustering data.* Prentice-Hall, Inc., NJ, USA, 1988.

[28] C. G. L. Jiang H. and Zhang X. C. Exclusive overall optimal solution of graph bipartition problem and backbone compute complexity. *Chinese Science Bulletin*, 52(17): 2077–2081, 2007.

[29] X. K. e. a. Jiang Y.X. and Tang C.J. Core-tag clustering for web2.0 based on multi-similarity measurements. In *The Joint International Conference on Asia-Pacific Web Conference (APWeb) and Web-Age Information Management (WAIM)*, pp. 222–233.

[30] M. I. Jordan and C. M. Bishop. "Neural Networks". *In Allen B. Tucker. Computer Science Handbook, Second Edition (Section VII: Intelligent Systems).* 2004.

[31] M. S. I. Karydis, A. Nanopoulos, H. -H. Gabriel and Myra. Tag-aware spectral clustering of music items, 2009.

[32] I. King and R. Baeza-Yates. *Weaving Services and People on the World Wide Web.* Springer, 2009.

[33] F. Lancaster. *Information Retrieval Systems: Characteristics, Testing and Evaluation.* Wiley, New York, 1968.

[34] A. H. Lashkari, F. Mahdavi and V. Ghomi. A boolean model in information retrieval for search engines, 2009.
[35] R. S. U. A. Lehwark, P. Visualization and clustering of tagged music data. data analysis. *Machine Learning and Applications*, pp. 673–680, 2008.
[36] L. F. e. a. Lei X. F., Xie K. Q. An efficient clustering algorithm based on local optimality of k-means. *Journal of Software*, 19(7): 1683–1692, 2008.
[37] U. V. Luxburg. A tutorial on spectral clustering. *Statistics and Computing*, 17(4), 2007.
[38] P. D. Martin Leginus and V. Zemaitis. Improving tensor based recommenders with clustering. *In: The 20th International Conference on User Modeling, Adaptation, and Personalization (UMAP'12)*, pp. 151–163. Springer-Verlag Berlin, Heidelberg.
[39] E. A. Matteo N. R., Peroni S. and Tamburini F. A parametric architecture for tags clustering in folksonomic search engines, 2009.
[40] G. H. Max Chevalier, Antonina Dattolo and E. Pitassi. Information retrieval and folksonomies together for recommender systems. *Systems E-Commerce and Web Technologies*, volume 85 of Lecture Notes: Chapter 15, pp. 172–183.
[41] T. J. H. W. S. A. M. L. Miao, G. Extracting data records from the web using tag path clustering. In *Proceedings of the 18th International Conference on World Wide Web*, pp. 981–990. ACM.
[42] H. F. Michael J. B. Technical comments comment on "Clustering by passing messages between data points". *Science*, 319: 726c–727c, 2008.
[43] P. Mika. Ontologies are us: A unified model of social networks and semantics. In Y. Gil, E. Motta, V. R. Benjamins and M. A. Musen, editors, *ISWC 2005, volume 3729 of LNCS, Berlin Heidelberg. Springer-Verlag.*, pp. 522–536, 2005.
[44] M. Montaner, B. Lopez and J. L. de la Rosa. A taxonomy of recommender agents on the internet. *Artificial Intelligence Review*, 19(4): 285?30, 2003.
[45] H. W. Ng T. and Raymond J. Clarans: A method for clustering objects for spatial data mining. *IEEE Transactions on Knowldge and Data Engineering*, 14(9): 1003–1026, 2002.
[46] F. T. e. a. Nicola R. D. and Silvio P. Of mice and terms: Clustering algorithms on ambiguous terms in folksonomies. In *The 2010 ACM symposium on Applied Computing SAC10*, pp. 844–848.
[47] B. F. P. D. Z. W. M. L. Rong Pan, Guandong Xu. Improving recommendations by the clustering of tag neighbours. *Journal of Convergence*, Section C, 3(1), 2012.
[48] S. Sen, S. K. Lam, A. M. Rashid, D. Cosley, D. Frankowski, J. Osterhouse, F. M. Harper and J. Riedl. tagging, communities, vocabulary, evolution, November 2006.
[49] D. Shakhnarovish and Indyk. Nearest-neighbor methods in learning and vision. *The MIT Press*, 2005.
[50] A. Shepitsen, J. Gemmell, B. Mobasher and R. Burke. Personalized recommendation in social tagging systems using hierarchical clustering. *In RecSys?008: Proceedings of the 2008 ACM conference on Recommender systems*, pp. 259–266, 2008.
[51] D. J. Su T. A deterministic method for initializing k-menas clustering, 2004.
[52] B. K. Teknomo. K-means clustering tutorial.
[53] K. Thearling. An introduction to data mining: Discovering hidden value in your data warehouse.
[54] R. D. Validity and A. K. Jain. Studies in clustering methodologies. *Pattern Recognition*, pp. 235–254, 1979.
[55] V. D. H. F. F. F. Van Dam, J. Searching and browsing tagspaces using the semantic tag clustering search framework. *In: S. Computing(ICSC), editor, 2010 IEEE Fourth International Conference*, pp. 436–439. IEEE.
[56] wiki/Recommender system. http://en.wikipedia.org/wiki/recommender system.
[57] M. -L. Zhang and Z. -H. Zhou. Ml-knn: A lazy learning approach to multi-label learning. *Pattern Recognition*, 40(7): 2038–2048, 2007.
[58] Q. L. e. a. Zhou J. L., Nie X.J. Web clustering based on tag set similarity. *Journal of Computers*, 6(1): 59–66, 2011.

[59] Y. Zong, G. Xu, P. Jin, Y. Zhang, E. Chen and R. Pan. APPECT: *An Approximate Backbone-Based Clustering Algorithm for Tags Advanced Data Mining and Applications,* volume 7120 of *Lecture Notes in Computer Science,* pp. 175–189. Springer Berlin/Heidelberg, 2011.

[60] L. M. C. Zong Y. and Jiang H. Approximate backbone guided reduction clustering algorithm. *Journal of electronics and information technology,* 31(2)(2953–2957), 2009.

[61] C. G. Zou P. and ZHou Z. H. Approximate backbone guided fast ant algorithm to qap. *Journal of Software,* 16(10): 1691–1698, 2005.

[62] Xiaoyuan Su and Taghi M. Khoshgoftaar. A survey of collaborative filtering techniques. *Advances in Artificial Intelligence,* Volume 2009, January 2009.

[63] Yehuda Koren. Factor in the neighbors: Scalable and accurate collaborative filtering. *ACM Transactions on Knowledge Discovery from Data (TKDD),* Volume 4, Issue 1, January 2010.

[64] Zhu, T., Greiner, R. and Haubl, G. A fuzzy hybrid collaborative filtering technique for web personalization. *WWW2003,* May, 2003, Budapest, Hungary.

[65] Claypool, M., Gokhale, A. and Miranda, T. Combining content-based and collaborative filters in an online newspaper. *ACM SIGIR Workshop on Recommender Systems.*

[66] Dasari Siva Krishna and K. Rajani Devi. Improving Accumulated Recommendation System A Comparative Study of Diversity Using Pagerank Algorithm Technique. *International Journal of Advanced Science and Technology.*

[67] Sergey Brin, Lawrence Page. The anatomy of a large-scale hypertextual web search engine. *Computer Networks.* Vol. 30, Issue 1-7, April 1, 1998.

[68] Kubatz, M., Gedikli, F., Jannach and D. LocalRank—Neighborhood-based, fast computation of tag recommendations. *12th International Conference on Electronic Commerce and Web Technologies - EC-Web 2011.*

[69] Christopher D. Manning, Prabhakar Raghavan, Hinrich Schutze. Introduction to Information Retrieval. Cambridge University Press.

[70] P. Tamayo, D. Slonim, J. Mesirov, Q. Zhu, S. Kitareewan, S. Dmitrovsky, E. Lander and T. R. Golub. Interpreting patterns of gene expression with self-organizing maps: methods and application to hematopoietic differentiation. *National Academy of Sciences.*

[71] Ulrike Von Luxburg. A Tutorial on Spectral Clustering. *Statistics and Computing.*

Index